T0320669

Mathematical
Olympiad
in China (2019–2020)

Problems and Solutions

Mathematical Olympiad Series

ISSN: 1793-8570

Series Editors: Lee Peng Yee *(Nanyang Technological University, Singapore)*
Xiong Bin *(East China Normal University, China)*

Published

Vol. 19 *Mathematical Olympiad in China (2019–2020):*
Problems and Solutions
edited by Bin Xiong (East China Normal University, China)

Vol. 18 *Mathematical Olympiad in China (2017–2018):*
Problems and Solutions
edited by Bin Xiong (East China Normal University, China)

Vol. 17 *Mathematical Olympiad in China (2015–2016):*
Problems and Solutions
edited by Bin Xiong (East China Normal University, China)

Vol. 16 *Sequences and Mathematical Induction:*
In Mathematical Olympiad and Competitions
Second Edition
by Zhigang Feng (Shanghai Senior High School, China)
translated by: Feng Ma, Youren Wang

Vol. 15 *Mathematical Olympiad in China (2011–2014):*
Problems and Solutions
edited by Bin Xiong (East China Normal University, China) &
Peng Yee Lee (Nanyang Technological University, Singapore)

Vol. 14 *Probability and Expectation*
by Zun Shan (Nanjing Normal University, China)
translated by: Shanping Wang (East China Normal University, China)

Vol. 13 *Combinatorial Extremization*
by Yuefeng Feng (Shenzhen Senior High School, China)

Vol. 12 *Geometric Inequalities*
by Gangsong Leng (Shanghai University, China)
translated by: Yongming Liu (East China Normal University, China)

The complete list of the published volumes in the series can be found at
http://www.worldscientific.com/series/mos

Vol. 19 | Mathematical Olympiad Series

Mathematical lympiad
in China (2019–2020)
Problems and Solutions

Editor-in-Chief
Xiong Bin
East China Normal University, China

English Translators
Zhao Wei
East China Normal University, China
Zhou Tianyou
Shanghai High School, China

Copy Editors
Ni Ming
Kong Lingzhi
Wan Yuanlin
East China Normal University Press, China

 East China Normal University Press

 World Scientific

Published by

East China Normal University Press
3663 North Zhongshan Road
Shanghai 200062
China

and

World Scientific Publishing Co. Pte. Ltd.
5 Toh Tuck Link, Singapore 596224
USA office: 27 Warren Street, Suite 401-402, Hackensack, NJ 07601
UK office: 57 Shelton Street, Covent Garden, London WC2H 9HE

Library of Congress Cataloging-in-Publication Data
Names: Xiong, Bin, editor.
Title: Mathematical Olympiad in China (2019–2020) : problems and solutions /
     editor Xiong Bin, East China Normal University, China.
Description: Shanghai, China : East China Normal University Press ; Hackensack, NJ :
     World Scientific, [2023] | Series: Mathematical olympiad series, 1793-8570 ; vol. 19
Identifiers: LCCN 2022013715 | ISBN 9789811256325 (hardcover) |
     ISBN 9789811257391 (paperback) | ISBN 9789811256332 (ebook) |
     ISBN 9789811256349 (ebook other)
Subjects: LCSH: International Mathematical Olympiad. | Mathematics--Problems, exercises, etc. |
     Mathematics--Competitions--China.
Classification: LCC QA43 .M31455 2023 | DDC 510.76--dc23/eng/20220330
LC record available at https://lccn.loc.gov/2022013715

British Library Cataloguing-in-Publication Data
A catalogue record for this book is available from the British Library.

Copyright © 2023 by East China Normal University Press and
World Scientific Publishing Co. Pte. Ltd.

All rights reserved. This book, or parts thereof, may not be reproduced in any form or by any means, electronic or mechanical, including photocopying, recording or any information storage and retrieval system now known or to be invented, without written permission from the Publisher.

For photocopying of material in this volume, please pay a copying fee through the Copyright Clearance Center, Inc., 222 Rosewood Drive, Danvers, MA 01923, USA. In this case permission to photocopy is not required from the publisher.

For any available supplementary material, please visit
https://www.worldscientific.com/worldscibooks/10.1142/12837#t=suppl

Typeset by Stallion Press
Email: enquiries@stallionpress.com

Printed in Singapore

Preface

The first time China participated in IMO was in 1985, when two students were sent to the 26th IMO. Since 1986, China has a team of 6 students at every IMO except in 1998 when it was held in Taiwan. So far, up to 2018, China has achieved the number one ranking in team effort 24 times. A great majority of students received gold medals. The fact that China obtained such encouraging results is due to, on one hand, Chinese students' hard work and perseverance, and on the other hand, the effort of the teachers in schools and the training offered by national coaches. We believe this is also a result of the education system in China, in particular, the emphasis on training of the basic skills in science education.

The materials of this book come from two volumes (vol. 2019 and vol. 2020) of a book series in Chinese "走向 IMO: 数学奥林匹克试题集锦" (*Forward to IMO: A Collection of Mathematical Olympiad Problems*). It is a collection of problems and solutions of the major mathematical competitions in China. It provides a glimpse of how the China national team is selected and formed. First, there is the China Mathematical Competition, a national event. It is held on the second Sunday of September every year. Through the competition, about 380 students are selected to join the China Mathematical Olympiad (commonly known as the winter camp), or in short CMO, in November. CMO lasts for five days. Both the type and the difficulty of the problems match those of IMO. Similarly, students are given three problems to solve in 4.5 hours each day. From CMO, 60 students are selected to form a national training team. The training takes place for two weeks in the month of March. After four to six tests, plus two qualifying examinations, six students are finally selected to form the national team, taking part in IMO in July of that year.

In view of the differences in education, culture and economy of the western part of China with the coastal part in eastern China, mathematical competitions in West China did not develop as fast as the rest of the country. In order to promote the activity of mathematical competition, and to enhance the level of mathematical competition, starting from 2001, China Mathematical Olympiad Committee organizes the China Western Mathematical Olympiad.

Since 2012, the China Western Mathematical Olympiad has been renamed the China Western Mathematical Invitation. The competition dates have been changed from the first half of October to the middle of August since 2013.

The development of this competition reignited the enthusiasm of Western students for mathematics. Once again, the figures of Western students often appeared in the national team.

Since 1995, there was no female student in the Chinese national team. In order to encourage more female students participating in the mathematical competition, starting from 2002, China Mathematical Olympiad Committee has been conducting the China Girls' Mathematical Olympiad. Again, the top twelve winners will be admitted directly into the CMO.

The authors of this book are coaches of the China national team. They are Xiong Bin, Wu Jianping, Leng Gangsong, Yu Hongbing, Yao Yijun, Qu Zhenhua, Li Ting, Xiao Liang, Wang Xinmao, Ai Yinhua, Wang Bin, Fu Yunhao, He Yijie, Zhang Sihui, and Lin Tianqi. Those who took part in the translation work are Zhao Wei and Zhou Tianyou. We are grateful to Qiu Zonghu, Wang Jie, Zhou Qin, Wu Jianping, and Pan Chengbiao for their guidance and assistance to the authors. We are grateful to Ni Ming, Kong Linzhi of East China Normal University Press. Their effort has helped make our job easier. We are also grateful to Zhang Ji of World Scientific Publishing for her hard work leading to the final publication of the book.

Authors

October 2021

Introduction

Early days

The International Mathematical Olympiad (IMO), founded in 1959, is one of the most competitive and highly intellectual activities in the world for high school students.

Even before IMO, there were already many countries which had mathematics competition. They were mainly the countries in Eastern Europe and in Asia. In addition to the popularization of mathematics and the convergence in educational systems among different countries, the success of mathematical competitions at the national level provided a foundation for the setting-up of IMO. The countries that asserted great influence are Hungary, the former Soviet Union, and the United States. Here is a brief history of the IMO and mathematical competition in China.

In 1894, the Department of Education in Hungary passed a motion and decided to conduct a mathematical competition for the secondary schools. The well-known scientist, *J. von Etövös*, was the Minister of Education at that time. His support in the event had made it a success and thus it was well publicized. In addition, the success of his son, *R. von Etövös*, who was also a physicist, in proving the principle of equivalence of the general theory of relativity by *A. Einstein* through experiment, had brought Hungary to the world stage in science. Thereafter, the prize for mathematics competition in Hungary was named "*Etövös* prize". This was the first formally organized mathematical competition in the world. In what follows, Hungary had indeed produced a lot of well-known scientists including *L. Fejér*, *G. Szegö*, *T. Radó*, *A. Haar* and *M. Riesz* (in real analysis), *D. König* (in combinatorics), *T. von Kármán* (in aerodynamics), and *J. C. Harsanyi* (in game theory), who had also won the Nobel Prize for Economics in 1994. They all were the winners of Hungary mathematical competition.

The top scientific genius of Hungary, *J. von Neumann*, was one of the leading mathematicians in the 20th century. *Neumann* was overseas while the competition took place. Later he did the competition himself and it took him half an hour to complete. Another mathematician worth mentioning is the highly productive number theorist *P. Erdös*. He was a pupil of *Fejér* and a winner of the Wolf Prize. *Erdös* was very passionate about mathematical competition and setting competition questions. His contribution to discrete mathematics was unique and greatly significant. The rapid progress and development of discrete mathematics over the subsequent decades had indirectly influenced the types of questions set in IMO. An internationally recognized prize was named after *Erdös* to honour those who had contributed to the education of mathematical competition. Professor *Qiu Zonghu* from China had won the prize in 1993.

In 1934, a famous mathematician *B. Delone* conducted a mathematical competition for high school students in Leningrad (now St. Petersburg). In 1935, Moscow also started organizing such events. Other than being interrupted during the World War II, these events had been carried on until today. As for the Russian Mathematical Competition (later renamed as the Soviet Mathematical Competition), it was not started until 1961. Thus, the former Soviet Union and Russia became the leading powers of Mathematical Olympiad. A lot of grandmasters in mathematics including the great *A. N. Kolmogorov* were all very enthusiastic about the mathematical competition. They would personally involve themselves in setting the questions for the competition. The former Soviet Union even called it the Mathematical Olympiad, believing that mathematics is the "gymnastics of thinking". These points of view gave a great impact on the educational community. The winner of the Fields Medal in 1998, *M. Kontsevich*, was once the first runner-up of the Russian Mathematical Competition. *G. Kasparov*, the international chess grandmaster, was once the second runner-up. *Grigori Perelman*, the winner of the Fields Medal in 2006 (but he declined), who solved the Poincaré's Conjecture, was a gold medalist of IMO in 1982.

In the United States of America, due to the active promotion by the renowned mathematician *G. D. Birkhoff* and his son, together with *G. Pólya*, the Putnam mathematics competition was organized in 1938 for junior undergraduates. Many of the questions were within the scope of high school students. The top five contestants of the Putnam mathematical competition would be entitled to the membership of Putnam. Many of these were eventually outstanding mathematicians. There were the famous *R. Feynman* (winner of the Nobel Prize for Physics, 1965), *K. Wilson* (winner of the Nobel Prize for Physics, 1982), *J. Milnor* (winner of the Fields Medal, 1962), *D. Mumford* (winner of the Fields Medal, 1974), and *D. Quillen* (winner of the Fields Medal, 1978).

Since 1972, in order to prepare for the IMO, the United States of America Mathematical Olympiad (USAMO) was organized. The standard of questions posed was very high, parallel to that of the Winter Camp in China. Prior to this, the United States had organized American High School Mathematics Examination (AHSME) for the high school students since 1950. This was at the junior level and yet the most popular mathematics competition in America. Originally, it was planned to select about 100 contestants from AHSME to participate in USAMO. However, due to the discrepancy in the level of difficulty between the two competitions and other restrictions, from 1983 onwards, an intermediate level of competition, namely, American Invitational Mathematics Examination (AIME), was introduced. Henceforth both AHSME and AIME became internationally well-known. Since 2000, AHSME was replaced by AMC 12 and AMC 10. Students who perform well on the AMC 12 and AMC 10 are invited to participate in AIME. The combined scores of the AMC 12 and the AIME are used to determine approximately 270 individuals that will be invited back to take the USAMO, while the combined scores of the AMC 10 and the AIME are used to determine approximately 230 individuals that will be invited to take the USAJMO (United States of America Junior Mathematical Olympiad), which started in 2010 and follows the same format as the USAMO. A few cities in China had participated in the competition and the results were encouraging.

Similarly, as in the former Soviet Union, the Mathematical Olympiad education was widely recognized in America. The book "How to Solve it" written by *George Polya* along with many other titles had been translated into many different languages. *George Polya* provided a whole series of general heuristics for solving problems of all kinds. His influence in the educational community in China should not be underestimated.

International Mathematical Olympiad

In 1956, the East European countries and the Soviet Union took the initiative to organize the IMO formally. The first International Mathematical Olympiad (IMO) was held in Brasov, Romania, in 1959. At that time, there were only seven participating countries, namely, Romania, Bulgaria, Poland, Hungary, Czechoslovakia, East Germany, and the Soviet Union.

Subsequently, the United States of America, United Kingdom, France, Germany, and also other countries including those from Asia joined. Today, the IMO had managed to reach almost all the developed and developing countries. Except in the year 1980 due to financial difficulties faced by the host country, Mongolia, there were already 59 Olympiads held and 107 countries and regions participating.

The mathematical topics in the IMO include algebra, combinatorics, geometry, number theory. These areas have provided guidance for setting questions for the competitions. Other than the first few Olympiads, each IMO is normally held in mid-July every year and the test paper consists of 6 questions in all. The actual competition lasts for 2 days for a total of 9 hours where participants are required to complete 3 questions each day. Each question is 7 points which total up to 42 points. The full score for a team is 252 marks. About half of the participants will be awarded a medal, where 1/12 will be awarded a gold medal. The numbers of gold, silver and bronze medals awarded are in the ratio of 1:2:3 approximately. In the case when a participant provides a better solution than the official answer, a special award is given.

Each participating country and region will take turn to host the IMO. The cost is borne by the host country. China had successfully hosted the 31st IMO in Beijing. The event had made a great impact on the mathematical community in China. According to the rules and regulations of the IMO, all participating countries are required to send a delegation consisting of a leader, a deputy leader and 6 contestants. The problems are contributed by the participating countries and are later selected carefully by the host country for submission to the international jury set up by the host country. Eventually, only 6 problems will be accepted for use in the competition. The host country does not provide any question. The short-listed problems are subsequently translated, if necessary, in English, French, German, Spain, Russian, and other working languages. After that, the team leaders will translate the problems into their own languages.

The answer scripts of each participating team will be marked by the team leader and the deputy leader. The team leader will later present the scripts of their contestants to the coordinators for assessment. If there is any dispute, the matter will be settled by the jury. The jury is formed by the various team leaders and an appointed chairman by the host country. The jury is responsible for deciding the final 6 problems for the competition. Their duties also include finalizing the grading standard, ensuring the accuracy of the translation of the problems, standardizing replies to written queries raised by participants during the competition, synchronizing differences in grading between the team leaders and the coordinators and also deciding on the cut-off points for the medals depending on the contestants' results as the difficulties of problems each year are different.

China had participated informally in the 26th IMO in 1985. Only two students were sent. Starting from 1986, except in 1998 when the IMO was held in Taiwan, China had always sent 6 official contestants to the IMO. Today, the Chinese contestants not only performed outstandingly in the IMO, but also in the International Physics, Chemistry, Informatics, and Biology Olympiads. This can be regarded as an indication that China pays great attention to the training of basic skills in mathematics and science education.

Winners of the IMO

Among all the IMO medalists, there were many of them who eventually became great mathematicians. They were also awarded the Fields Medal, Wolf Prize and Nevanlinna Prize (a prominent mathematics prize for computing and informatics). In what follows, we name some of the winners.

G. Margulis, a silver medalist of IMO in 1959, was awarded the Fields Medal in 1978. *L. Lovasz*, who won the Wolf Prize in 1999, was awarded the Special Award in IMO consecutively in 1965 and 1966. *V. Drinfeld*, a gold medalist of IMO in 1969, was awarded the Fields Medal in 1990. *J.-C. Yoccoz* and *T. Gowers*, who were both awarded the Fields Medal in 1998, were gold medalists in IMO in 1974 and 1981 respectively. A silver medalist of IMO in 1985, *L. Lafforgue*, won the Fields Medal in 2002. A gold medalist of IMO in 1982, *Grigori Perelman* from Russia, was awarded the Fields Medal in 2006 for solving the final step of the Poincaré conjecture. In 1986, 1987, and 1988, *Terence Tao* won a bronze, silver, and gold medal respectively. He was the youngest participant to date in the IMO, first competing at the age of ten. He was also awarded the Fields Medal in 2006. Gold medalist of IMO 1988 and 1989, *Ngo Bau Chao*, won the Fields Medal in 2010, together with the bronze medalist of IMO 1988, *E. Lindenstrauss*. Gold medalist of IMO 1994 and 1995, *Maryam Mirzakhani* won the Fields Medal in 2014. A gold medalist of IMO in 1995, Artur Avila won the Fields Medal in 2014. Gold medalist of IMO 2005,2006 and 2007, Peter Scholze won the Fields Medal in 2018. A Bronze medalist of IMO in 1994, Akshay Venkatesh won the Fields Medal in 2018.

A silver medalist of IMO in 1977, *P. Shor*, was awarded the Nevanlinna Prize. A gold medalist of IMO in 1979, *A. Razborov*, was awarded the

Nevanlinna Prize. Another gold medalist of IMO in 1986, *S. Smirnov*, was awarded the Clay Research Award. *V. Lafforgue*, a gold medalist of IMO in 1990, was awarded the European Mathematical Society prize. He is *L. Lafforgue*'s younger brother.

Also, a famous mathematician in number theory, *N. Elkies*, who is also a professor at Harvard University, was awarded a gold medal of IMO in 1982. Other winners include *P. Kronheimer* awarded a silver medal in 1981 and *R. Taylor* a contestant of IMO in 1980.

Mathematical competition in China

Due to various reasons, mathematical competition in China started relatively late but is progressing vigorously.

"We are going to have our own mathematical competition too!" said *Hua Luogeng*. *Hua* is a household name in China. The first mathematical competition was held concurrently in Beijing, Tianjin, Shanghai, and Wuhan in 1956. Due to the political situation at the time, this event was interrupted a few times. Until 1962, when the political environment started to improve, Beijing and other cities started organizing the competition though not regularly. In the era of Cultural Revolution, the whole educational system in China was in chaos. The mathematical competition came to a complete halt. In contrast, the mathematical competition in the former Soviet Union was still on-going during the war and at a time under the difficult political situation. The competitions in Moscow were interrupted only 3 times between 1942 and 1944. It was indeed commendable.

In 1978, it was the spring of science. *Hua Luogeng* conducted the Middle School Mathematical Competition for 8 provinces in China. The mathematical competition in China was then making a fresh start and embarked on a road of rapid development. *Hua* passed away in 1985. In commemorating him, a competition named *Hua Luogeng* Gold Cup was set up in 1986 for students in Grade 6 and 7 and it has a great impact.

The mathematical competitions in China before 1980 can be considered as the initial period. The problems set were within the scope of middle school textbooks. After 1980, the competitions were gradually moving towards the senior middle school level. In 1981, the Chinese Mathematical Society decided to conduct the China Mathematical Competition, a national event for high schools.

In 1981, the United States of America, the host country of IMO, issued an invitation to China to participate in the event. Only in 1985, China sent two contestants to participate informally in the IMO. The results were not encouraging. In view of this, another activity called the Winter Camp was conducted after the China Mathematical Competition. The Winter Camp was later renamed as the China Mathematical Olympiad or CMO. The winning team would be awarded the *Chern Shiing-Shen* Cup. Based on the outcome at the Winter Camp, a selection would be made to form the 6-member national team for IMO. From 1986 onwards, other than the year when IMO was organized in Taiwan, China had been sending a 6-member team to IMO. Up to 2018, China had been awarded the overall team champion for 19 times.

In 1990, China had successfully hosted the 31st IMO. It showed that the standard of mathematical competition in China has leveled that of other leading countries. First, the fact that China achieves the highest marks at the 31st IMO for the team is evidence of the effectiveness of the pyramid approach in selecting the contestants in China. Secondly, the Chinese mathematicians had simplified and modified over 100 problems and submitted them to the team leaders of the 35 countries for their perusal. Eventually, 28 problems were recommended. At the end, 5 problems were chosen (IMO requires 6 problems). This is another evidence to show that China has achieved the highest quality in setting problems. Thirdly, the answer scripts of the participants were marked by the various team leaders and assessed by the coordinators who were nominated by the host countries. China had formed a group 50 mathematicians to serve as coordinators who would ensure the high accuracy and fairness in marking. The marking process was completed half a day earlier than it was scheduled. Fourthly, that was the first ever IMO organized in Asia. The outstanding performance by China had encouraged the other developing countries, especially those in Asia. The organizing and coordinating work of the IMO by the host country was also reasonably good.

In China, the outstanding performance in mathematical competition is a result of many contributions from all quarters of mathematical community. There are the older generation of mathematicians, middle-aged mathematicians and also the middle and elementary school teachers. There is one person who deserves a special mention, and he is *Hua Luogeng*. He initiated and promoted the mathematical competition. He is also the author of the following books: Beyond *Yang hui*'s Triangle, Beyond the *pi* of *Zu Chongzhi*, Beyond the Magic Computation of *Sun-zi*, Mathematical Induction, and Mathematical Problems of Bee Hive. These were his books derived from mathematics competitions. When China resumed mathematical competition in 1978, he participated in setting problems and giving critique to solutions of the problems. Other outstanding books derived from the Chinese mathematics competitions are: Symmetry by *Duan Xuefu*, Lattice and Area by *Min Sihe*, One Stroke Drawing and Postman Problem by *Jiang Boju*.

After 1980, the younger mathematicians in China had taken over from the older generation of mathematicians in running the mathematical competition. They worked and strived hard to bring the level of mathematical competition in China to a new height. *Qiu Zonghu* is one such outstanding representative. From the training of contestants and leading the team 3 times to IMO to the organizing of the 31st IMO in China, he had contributed prominently and was awarded the *P. Erdös* prize.

Preparation for IMO

Currently, the selection process of participants for IMO in China is as follows.

First, the China Mathematical Competition, a national competition for high Schools, is organized on the second Sunday in September every year. The objectives are: to increase the interest of students in learning mathematics, to promote the development of co-curricular activities in mathematics, to help improve the teaching of mathematics in high schools, to discover and cultivate the talents and also to prepare for the IMO. This happens since 1981. Currently there are about 500,000 participants taking part.

Through the China Mathematical Competition, around 350 of students are selected to take part in the China Mathematical Olympiad or CMO, that is, the Winter Camp. The CMO lasts for 5 days and is held in November every year. The types and difficulties of the problems in CMO are very much similar to the IMO. There are also 3 problems to be completed within 4.5 hours each day. However, the score for each problem is 21 marks which add up to 126 marks in total. Starting from 1990, the Winter Camp instituted the *Chern Shiing-Shen* Cup for team championship. In 1991, the Winter Camp was officially renamed as the China Mathematical Olympiad (CMO). It is similar to the highest national mathematical competition in the former Soviet Union and the United States.

The CMO awards the first, second and third prizes. Among the participants of CMO, about 60 students are selected to participate in the training for IMO. The training takes place in March every year. After 6 to 8 tests and another 2 rounds of qualifying examinations, only 6 contestants are short-listed to form the China IMO national team to take part in the IMO in July.

Besides the China Mathematical Competition (for high schools), the Junior Middle School Mathematical Competition is also developing well. Starting from 1984, the competition is organized in April every year by the Popularization Committee of the Chinese Mathematical Society. The various provinces, cities and autonomous regions would rotate to host the event. Another mathematical competition for the junior middle schools is also conducted in April every year by the Middle School Mathematics Education Society of the Chinese Educational Society since 1998 till now.

The *Hua Luogeng* Gold Cup, a competition by invitation, had also been successfully conducted since 1986. The participating students comprise elementary six and junior middle one students. The format of the competition consists of a preliminary round, semi-finals in various provinces, cities, and autonomous regions, then the finals.

Mathematical competition in China provides a platform for students to showcase their talents in mathematics. It encourages learning of mathematics among students. It helps identify talented students and to provide them with differentiated learning opportunity. It develops co-curricular activities in mathematics. Finally, it brings about changes in the teaching of mathematics.

Contents

China Mathematical Competition

2018

While the scope of the test questions in the first round of the 2018 China Mathematical Competition does not exceed the teaching requirements and content specified in the "General High School Mathematics Curriculum Standards (Experiments)" promulgated by the Ministry of Education of China in 2003, the methods of proposing the questions have been improved. The emphasis placed on to test the students' basic knowledge and skills, and their abilities to integrate and use flexibly of them. Each test paper includes eight fill-in-the-blank questions and three answer questions. The answer time is 80 minutes, and the full score is 120 points.

The scope of the test questions in the second round (Complementary Test) is in line with the International Mathematical Olympiad, with some expanded knowledge, plus a few contents of the Mathematical Competition Syllabus. Each test paper consists of four answer questions, including a plane geometry one, and the answering time is 150 minutes. The full score is 180 points.

Test Paper A
(8:10 – 9:20; September 9, 2018)

Part I Short-Answer Questions (Questions 1–8, eight marks each)

1 Given $A = \{1, 2, 3, \ldots, 99\}$, $B = \{2x \mid x \in A\}$, $C = \{x \mid 2x \in A\}$, the number of elements of $B \cap C$ is _____.

Solution By the given condition, we have

$$B \cap C = \{2, 4, 6, \ldots, 198\} \cap \left\{ \frac{1}{2}, 1, \frac{3}{2}, 2, \ldots, \frac{99}{2} \right\}$$

$$= \{2, 4, 6, \ldots, 48\}.$$

Consequently, the number of elements of $B \cap C$ is 24. $\square$

2 Let the distance from point P to plane α be 3 and point Q be on α such that the angle formed by line PQ and α is between $30°$ and $60°$. The area of the region formed by such a point Q is _____.

Solution Let the projection of point P on plane α be O. By the given condition we have

$$\frac{OP}{OQ} = \tan \angle OQP \in \left[\frac{\sqrt{3}}{3}, \sqrt{3} \right],$$

i.e., $OQ \in [1, 3]$. Therefore, the area of the region desired is $\pi \cdot 3^2 - \pi \cdot 1^2 = 8\pi$. $\square$

3 Let 1, 2, 3, 4, 5, 6 be randomly arranged in a line, denoted as a, b, c, d, e, f. Then the probability of $abc + def$ being even numbers is _____.

Solution Firstly consider the case that $abc + def$ is odd. Then one of abc, def is odd and the other is even. If abc is odd, then a, b, c are permutations of $1, 3, 5$, and hence d, e, f are permutations of $2, 4, 6$. There are $3! \times 3! = 36$ such cases. By symmetry, the number of cases that $abc + def$ is odd is $36 \times 2 = 72$.

Therefore, the probability of $abc + def$ being even is $1 - \dfrac{72}{6!} = 1 - \dfrac{72}{720} = \dfrac{9}{10}$. $\square$

4 In plane rectangular coordinate system xOy, the left and right foci of ellipse C: $\dfrac{x^2}{a^2} + \dfrac{y^2}{b^2} = 1 \, (a > b > 0)$ are F_1, F_2, respectively. Chords ST and UV of C are parallel to x and y axes, respectively, and intersect at point P. The lengths of line segments PU, PS, PV, PT are 1, 2, 3, 6, respectively. The area of $\triangle PF_1F_2$ is _____.

Solution By symmetry, without loss of generality, assume $P(x_P, y_P)$ is in the first quadrant. By the given condition we find

$$x_P = \frac{1}{2}(|PT| - |PS|) = 2, y_P = \frac{1}{2}(|PV| - |PU|) = 1,$$

and thus $P(2,1)$. From $x_P = |PU| = 1, |PS| = 2$, we get $U(2,2), S(4,1)$. Substituting them into the equation of ellipse C, we have

$$4 \cdot \frac{1}{a^2} + 4 \cdot \frac{1}{b^2} = 16 \cdot \frac{1}{a^2} + \frac{1}{b^2} = 1,$$

and the solution is $a^2 = 20, b^2 = 5$.

Consequently, $S_{\triangle PF_1F_2} = \dfrac{1}{2} \cdot |F_1F_2| \cdot |y_P| = \sqrt{a^2 - b^2} \cdot y_P = \sqrt{15}$. $\quad\square$

5 Suppose $f(x)$ is an even periodic function with period 2 defined on $\mathbb{R}$ and is strictly decreasing on interval $[0,1]$, satisfying $f(\pi) = 1, f(2\pi) = 2$. Then the solution set of the system of inequalities $\begin{cases} 1 \le x \le 2, \\ 1 \le f(x) \le 2 \end{cases}$ is _____.

Solution From the given condition that $f(x)$ is an even function and is strictly decreasing on interval $[0,1]$, we learn that $f(x)$ is strictly increasing on the interval $[-1,0]$. Combined with the fact that $f(x)$ is a periodic function with period 2, it follows that $[1,2]$ is a strictly increasing interval of $f(x)$.

Notice that

$$f(\pi - 2) = f(\pi) = 1, f(8 - 2\pi) = f(-2\pi) = f(2\pi) = 2,$$

and thus

$$1 \le f(x) \le 2 \Leftrightarrow f(\pi - 2) \le f(x) \le f(8 - 2\pi).$$

Then we have $1 < \pi - 2 < 8 - 2\pi < 2$. Therefore, the original system of inequalities is valid if and only if $x \in [\pi - 2, 8 - 2\pi]$. $\quad\square$

6 Suppose complex number z satisfies $|z| = 1$ such that the equation $zx^2 + 2\overline{z}x + 2 = 0$ with respect to x has a real root. Then the sum of such complex numbers z is _____.

Solution   Let $z = a + b\mathrm{i}$ ($a, b \in \mathbb{R}, a^2 + b^2 = 1$). The original equation can be rewritten as

$$(a + b\mathrm{i})x^2 + 2(a - b\mathrm{i})x + 2 = 0.$$

Separating the real and imaginary parts of the above equation, we find,

$$ax^2 + 2ax + 2 = 0, \qquad\qquad ①$$

$$bx^2 - 2bx = 0. \qquad\qquad ②$$

If $b = 0$, then $a^2 = 1$. Since Equation ① has no real number solution for $a = 1$, then $a = -1$, and there exists real numbers $x = -1 \pm \sqrt{3}$ satisfying Equations ① and ②. So $z = -1$ satisfies the condition.

If $b \neq 0$, then from Equation ② we learn that $x \in \{0, 2\}$. Obviously, $x = 0$ does not satisfy Equation ①. Therefore, $x = 2$. Substituting it into Equation ①, we get the solution $a = -\dfrac{1}{4}$, and $b = \pm\dfrac{\sqrt{15}}{4}$. Then $z = \dfrac{-1 \pm \sqrt{15}\mathrm{i}}{4}$.

To sum up, the sum of all the complex numbers z satisfying the condition is

$$-1 + \frac{-1 + \sqrt{15}\mathrm{i}}{4} + \frac{-1 - \sqrt{15}\mathrm{i}}{4} = -\frac{3}{2}. \qquad\qquad \square$$

7 Suppose O is the excentre of $\triangle ABC$. If $\overrightarrow{AO} = \overrightarrow{AB} + 2\overrightarrow{AC}$, then the value of $\sin \angle BAC$ is _____.

Solution   Without loss of generality, let the circumscribed circle of $\triangle ABC$ have radius $R = 2$. By the given condition, we have

$$2\overrightarrow{AC} = \overrightarrow{AO} - \overrightarrow{AB} = \overrightarrow{BO}, \qquad\qquad ①$$

and hence $AC = \dfrac{1}{2}BO = 1$.

Taking the midpoint M of AC, it follows that $OM \perp AC$. By Equation ①, we know that $OM \perp BO$ and that A and B lie on the same side of line OM. Therefore,

$$\cos \angle BOC = \cos(90° + \angle MOC) = -\sin \angle MOC = -\frac{MC}{OC} = -\frac{1}{4}.$$

In $\triangle BOC$, the law of cosines gives

$$BC = \sqrt{OB^2 + OC^2 - 2OB \cdot OC \cdot \cos \angle BOC} = \sqrt{10}.$$

Furthermore, in $\triangle ABC$ by the law of sines we get $\sin \angle BAC = \dfrac{BC}{2R} = \dfrac{\sqrt{10}}{4}$. $\qquad\square$

8 Suppose integer sequence $a_1, a_2, \ldots, a_{10}$ satisfies $a_{10} = 3a_1, a_2 + a_8 = 2a_5$, and

$$a_{i+1} \in \{1 + a_i, 2 + a_i\}, \ i = 1, 2, \ldots, 9.$$

Then the number of such sequences is _____.

Solution Let $b_i = a_{i+1} - a_i \in \{1, 2\}(i = 1, 2, \ldots, 9)$. It follows that

$$2a_1 = a_{10} - a_1 = b_1 + b_2 + \cdots + b_9, \qquad \text{①}$$

$$b_2 + b_3 + b_4 = a_5 - a_2 = a_8 - a_5 = b_5 + b_6 + b_7. \qquad \text{②}$$

Denote the numbers of terms in b_2, b_3, b_4 that have value 2 by t. By Equation ②, t is also the number of terms in b_5, b_6, b_7 with value 2, where $t \in \{0, 1, 2, 3\}$. Therefore, the number of value choices of $b_2, b_3, \ldots, b_7$ is

$$(C_3^0)^2 + (C_3^1)^2 + (C_3^2)^2 + (C_3^3)^2 = 20.$$

After taking the value of $b_2, b_3, \ldots, b_7$, the values of b_8, b_9 can be specified arbitrarily in $2^2 = 4$ ways.

Finally, by Equation ①, we should take $b_1 \in \{1, 2\}$ such that $b_1 + b_2 + \cdots + b_9$ is even. Since b_1 is taken uniquely, the value of integer a_1 is determined. Thus sequence $b_1, b_2, \ldots, b_9$ uniquely corresponds to sequence $a_1, a_2, \ldots, a_{10}$, satisfying the given condition.

In conclusion, the number of sequences that satisfy the given condition is $20 \times 4 = 80$. $\qquad\square$

Part II Word Problems (16 marks for Question 9, 20 marks each for Questions 10 and 11, and then 56 marks in total)

9 (16 marks) Given function $f(x) = \begin{cases} |\log_3 x - 1|, & 0 < x \leqslant 9, \\ 4 - \sqrt{x}, & x > 9. \end{cases}$ defined on $\mathbb{R}^+$, let a, b, c be three real numbers that are not equal to each other, satisfying $f(a) = f(b) = f(c)$. Find the range of abc.

Solution   Since $f(x)$ is strictly decreasing on $(0, 3]$, strictly increasing on $[3, 9]$ and strictly decreasing on $[9, +\infty)$, with $f(3) = 0, f(9) = 1$, therefore, from the graph of the function, we find

$$a \in (0, 3), b \in (3, 9), c \in (9, +\infty),$$

and also $f(a) = f(b) = f(c) \in (0, 1)$. As $f(a) = f(b)$, then

$$1 - \log_3 a = \log_3 b - 1.$$

Taking $\log_3 a + \log_3 b = 2$, then $ab = 3^2 = 9$. Thus $abc = 9c$.
Since

$$0 < f(c) = 4 - \sqrt{c} < 1,$$

thus $c \in (9, 16)$. Then $abc = 9c \in (81, 144)$.
Therefore, the range of abc is $(81, 144)$.   □

Remark   For any $r \in (81, 144)$, taking $c_0 = \dfrac{r}{9}$, then $c_0 \in (9, 16)$, and hence $f(c_0) \in (0, 1)$. Draw line l passing through point $(c_0, f(c_0))$ and being parallel to the x-axis, and then l intersects with the graph of $f(x)$ at two other points: $(a, f(a)), (b, f(b))$, where $a \in (0, 3), b \in (3, 9)$. $f(a) = f(b) = f(c)$ and $ab = 9$ are satisfied, and thus $abc = r$.

10 (20 marks) It is known that a sequence of real numbers $a_1, a_2, a_3, \ldots$ satisfies: for any positive integer n,

$$a_n(2S_n - a_n) = 1,$$

where S_n is the sum of the first n terms of the sequence. Prove that

(1) For any positive integer n, $a_n < 2\sqrt{n}$;
(2) For any positive integer n, $a_n a_{n+1} < 1$.

Solution   (1) We conventionally assume that $S_0 = 0$. By the condition, for any positive integer n, we have

$$1 = a_n(2S_n - a_n) = (S_n - S_{n-1})(S_n + S_{n-1}) = S_n^2 - S_{n-1}^2.$$

Then $S_n^2 = n + S_0^2 = n$, and that is $S_n = \pm\sqrt{n}$ (This also holds when $n = 0$).

Obviously, we find $a_n = S_n - S_{n-1} \leq \sqrt{n} + \sqrt{n-1} < 2\sqrt{n}$.

(2) It is only necessary to consider the case where a_n, a_{n+1} have the same sign. Without loss of generality, and we assume a_n, a_{n+1} are both positive (otherwise the condition would still be satisfied by turning each item of the sequence into the opposite together). Then $S_{n+1} > S_n > S_{n-1} > -\sqrt{n}$, so there must be

$$S_n = \sqrt{n}, S_{n+1} = \sqrt{n+1},$$

and then

$$a_n = \sqrt{n} - \sqrt{n-1}, \quad a_{n+1} = (\sqrt{n+1} - \sqrt{n}).$$

Therefore,

$$a_n a_{n+1} < (\sqrt{n} + \sqrt{n-1})(\sqrt{n+1} - \sqrt{n})$$
$$< (\sqrt{n+1} + \sqrt{n})(\sqrt{n+1} - \sqrt{n}) = 1. \qquad \square$$

11 (20 marks) In a plane rectangular coordinate system xOy, let AB be the chord of parabola $y^2 = 4x$ passing through point $F(1,0)$. The circumscribed circle of $\triangle AOB$ intersects the parabola at point P, which is different from O, A, B. If PF bisects $\angle APB$, find all the possible values of $|PF|$.

Solution Let $A\left(\dfrac{y_1^2}{4}, y_1\right), B\left(\dfrac{y_2^2}{4}, y_2\right), P\left(\dfrac{y_3^2}{4}, y_3\right)$. It is known from the given condition that y_1, y_2, y_3 are unequal to each other and non-zero.

Let the equation of line AB be $x = ty + 1$. Combing with the equation of the parabola we obtain

$$y^2 - 4ty - 4 = 0,$$

thus

$$y_1 y_2 = -4. \qquad \text{(1)}$$

It should be noticed the circumscribed circle of $\triangle AOB$ passes through point O, and we can set the equation of the circle as

$$x^2 + y^2 + dx + ey = 0.$$

Combing with $x = \dfrac{y^2}{4}$, it follows that

$$\frac{y^4}{16} + \left(1 + \frac{d}{4}\right)y^2 + ey = 0.$$

This quadratic equation has four different real roots $y = y_1, y_2, y_3, 0$. By Vieta's formulas we have $y_1 + y_2 + y_3 + 0 = 0$, and thus

$$y_3 = -(y_1 + y_2). \qquad \qquad ②$$

Since PF bisects $\angle APB$, by angle bisector theorem we have

$$\frac{|PA|}{|PB|} = \frac{|FA|}{|FB|} = \frac{|y_1|}{|y_2|},$$

Combing ① and ②, it follows that

$$\begin{aligned}
\frac{y_1^2}{y_2^2} = \frac{|PA|^2}{|PB|^2} &= \frac{\left(\dfrac{y_3^2}{4} - \dfrac{y_1^2}{4}\right)^2 + (y_3 - y_1)^2}{\left(\dfrac{y_3^2}{4} - \dfrac{y_2^2}{4}\right)^2 + (y_3 - y_2)^2} \\[2mm]
&= \frac{((y_1 + y_2)^2 - y_1^2)^2 + 16(2y_1 + y_2)^2}{((y_1 + y_2)^2 - y_2^2)^2 + 16(2y_2 + y_1)^2} \\[2mm]
&= \frac{(y_2^2 - 8)^2 + 16(4y_1^2 + y_2^2 - 16)}{(y_1^2 - 8)^2 + 16(4y_2^2 + y_1^2 - 16)} = \frac{y_2^4 + 64y_1^2 - 192}{y_1^4 + 64y_2^2 - 192},
\end{aligned}$$

namely, $y_1^6 + 64y_1^2 y_2^2 - 192y_1^2 = y_2^6 + 64y_2^2 y_1^2 - 192y_2^2$, and thus

$$(y_1^2 - y_2^2)(y_1^4 + y_1^2 y_2^2 + y_2^4 - 192) = 0.$$

When $y_1^2 = y_2^2$, $y_2 = -y_1$, therefore $y_3 = 0$. This means that P coincides with O, which is inconsistent with the given condition.

When $y_1^4 + y_1^2 y_2^2 + y_2^4 - 192 = 0$, together with Equation ①, we find

$$(y_1^2 + y_2^2)^2 = 192 + (y_1 y_2)^2 = 208.$$

Since $y_1^2 + y_2^2 = 4\sqrt{13} > 8 = |2y_1 y_2|$, therefore there exists real numbers y_1, y_2 satisfying Equation ① as well as $y_1^2 + y_2^2 = 4\sqrt{13}$, corresponding to the points A, B that satisfy the given condition. Furthermore, combing Equations ① and ② we obtain

$$\begin{aligned}
|PF| &= \frac{y_3^2}{4} + 1 = \frac{(y_1 + y_2)^2 + 4}{4} = \frac{y_1^2 + y_2^2 - 4}{4} \\[2mm]
&= \frac{\sqrt{208} - 4}{4} = \sqrt{13} - 1.
\end{aligned}$$

$\square$

Test Paper B
(8:00 – 9:20; September 9, 2018)

Part I Short-Answer Questions (Questions 1–8, eight marks each)

1. Given $A = \{2, 0, 1, 8\}$, $B = \{2a \mid a \in A\}$, then the sum of all the elements of $A \cup B$ is _____ .

Solution It is easy to find $B = \{4, 0, 2, 16\}$, and thus $A \cup B = \{0, 1, 2, 4, 8, 16\}$. Therefore, the sum of all the elements of $A \cup B$ is $0 + 1 + 2 + 4 + 8 + 16 = 31$. □

2. Given a cone with vertex P, base radius 2 and height 1, a point Q is taken on its base, such that the angle between line PQ and the base is not greater than $45°$. Then the area formed by such points Q's that satisfy the condition is _____ .

Solution The projection of vertex P on the base is the center of the base denoted by O. By the given condition we have $\dfrac{OP}{OQ} = \tan \angle OQP \leq 1$, namely, $OQ \geq 1$.

Therefore, the area of the region desired is $\pi \cdot 2^2 - \pi \cdot 1^2 = 3\pi$. □

3. Let 1, 2, 3, 4, 5, 6 are randomly arranged in a line, denoted as a, b, c, d, e, f. Then the probability of $abc + def$ being odds numbers is _____ .

Solution If $abc + def$ is odd, then one of abc, def is odd and the other is even. If abc is odd, then a, b, c are permutations of 1, 3, 5, and hence d, e, f are permutations of 2, 4, 6. There are $3! \times 3! = 36$ such cases. By symmetry, the number of cases that $abc + def$ is odd is $36 \times 2 = 72$.

Therefore, the probability of $abc + def$ being odd is $\dfrac{72}{6!} = \dfrac{72}{720} = \dfrac{1}{10}$. □

4. In a plane rectangular coordinate system xOy, line l passes through the origin and $\vec{n} = (3, 1)$ is a normal vector of l. Suppose sequence $\{a_n\}$ satisfies: for any positive integer n, point (a_{n+1}, a_n) is on l. If $a_2 = 6$, then the value of $a_1 a_2 a_3 a_4 a_5$ is _____ .

Solution It is easy to know that the equation of line l is $3x + y = 0$. Then for any positive integer n,

$$3a_{n+1} + a_n = 0,$$

namely, $a_{n+1} = -\dfrac{1}{3}a_n$. Therefore, $\{a_n\}$ is a geometric sequence with common ratio $-\dfrac{1}{3}$. So we get

$$a_3 = -\frac{1}{3}a_2 = -2.$$

From the properties of geometric sequence, it follows that

$$a_1 a_2 a_3 a_4 a_5 = a_3^5 = (-2)^5 = -32. \qquad \square$$

5 Suppose α, β satisfy $\tan\left(\alpha + \dfrac{\pi}{3}\right) = -3, \tan\left(\beta - \dfrac{\pi}{6}\right) = 5$. The value of $\tan(\alpha - \beta)$ is _____.

Solution Using the difference formula for tangent, we have

$$\tan\left(\left(\alpha + \frac{\pi}{3}\right) - \left(\beta - \frac{\pi}{6}\right)\right) = \frac{-3 - 5}{1 + (-3) \times 5} = \frac{4}{7},$$

namely, $\tan\left(\alpha - \beta + \dfrac{\pi}{2}\right) = \dfrac{4}{7}$. Therefore,

$$\tan(\alpha - \beta) = -\cot\left(\alpha - \beta + \frac{\pi}{2}\right) = -\frac{7}{4}. \qquad \square$$

6 Let the directrix of parabola C : $y^2 = 2x$ intersect the x-axis at point A. Make line l through point $B(-1, 0)$ and tangent to C at point K. Construct a parallel line to l through point A, intersecting C at points M, N. Then the area of $\triangle KMN$ is _____.

Solution Let the slope of the parallel lines l and MN be k. Then we have

$$l : x = \frac{1}{k}y - 1, \, MN : x = \frac{1}{k}y - \frac{1}{2}.$$

Combining l with C gives equation $y^2 - \dfrac{2}{k}y + 2 = 0$. By the given condition, its discriminant is zero. Therefore,

$$k = \pm\frac{\sqrt{2}}{2}.$$

Combining MN with C leads to equation $y^2 - \dfrac{2}{k}y + 1 = 0$, hence

$$|y_M - y_N| = \sqrt{(y_M + y_N)^2 - 4y_M y_N} = \sqrt{\dfrac{4}{k^2} - 4} = 2.$$

Since l is parallel to MN, then we have

$$S_{\triangle KMN} = S_{\triangle BMN} = |S_{\triangle BAM} - S_{\triangle BAN}|$$
$$= \dfrac{1}{2} \cdot |AB| \cdot |y_M - y_N|$$
$$= \dfrac{1}{2} \cdot \dfrac{1}{2} \cdot 2 = \dfrac{1}{2}. \qquad \square$$

7 Suppose $f(x)$ is an even periodic function with period 2 defined on $\mathbb{R}$ and is strictly decreasing on interval $[1, 2]$, satisfying $f(\pi) = 1, f(2\pi) = 0$. Then the solution set of the system of inequalities $\begin{cases} 0 \le x \le 1, \\ 0 \le f(x) \le 1 \end{cases}$ is _____.

Solution Since $f(x)$ is an even function and is strictly decreasing on interval $[1, 2]$, then $f(x)$ is strictly increasing on the interval $[-2, -1]$. Combining with the fact that $f(x)$ has a period of 2, it follows that $[0, 1]$ is a strictly increasing interval of $f(x)$. Notice that

$$f(4 - \pi) = f(\pi - 4) = f(\pi) = 1, f(2\pi - 6) = f(2\pi) = 0,$$

and thus

$$0 \le f(x) \le 1 \Leftrightarrow f(2\pi - 6) \le f(x) \le f(4 - \pi).$$

Since $0 < 2\pi - 6 < 4 - \pi < 1$, therefore the above system of inequalities is valid if and only if $x \in [2\pi - 6, 4 - \pi]$. $\qquad \square$

8 Suppose complex numbers z_1, z_2, z_3 satisfy $|z_1| = |z_2| = |z_3| = 1$, $|z_1 + z_2 + z_3| = r$, where r is a given real number. Then the real part of $\dfrac{z_1}{z_2} + \dfrac{z_2}{z_3} + \dfrac{z_3}{z_1}$ is _____ (Expressed by r).

Solution Denote $w = \dfrac{z_1}{z_2} + \dfrac{z_2}{z_3} + \dfrac{z_3}{z_1}$. From the properties of complex moduli, it follows that

$$\overline{z_1} = \dfrac{1}{z_1}, \overline{z_2} = \dfrac{1}{z_2}, \overline{z_3} = \dfrac{1}{z_3}.$$

Hence $w = z_1\overline{z_2} + z_2\overline{z_3} + z_3\overline{z_1}$. We have

$$r^2 = (z_1 + z_2 + z_3)(\overline{z}_1 + \overline{z}_2 + \overline{z}_3)$$
$$= |z_1|^2 + |z_2|^2 + |z_3|^2 + w + \overline{w} = 3 + 2\operatorname{Re} w.$$

Therefore, the solution is $\operatorname{Re} w = \dfrac{r^2 - 3}{2}$. □

Part II Word Problems (16 marks for Question 9, 20 marks each for Questions 10 and 11, and then 56 marks in total)

9 (16 marks) Given sequence $\{a_n\}$: $a_1 = 7, \dfrac{a_{n+1}}{a} = a_n + 2, n = 1, 2, 3, \ldots$, find the smallest positive integer n that satisfies $a_n > 4^{2018}$.

Solution   From $\dfrac{a_{n+1}}{a_n} = a_n + 2$, we have $a_{n+1} + 1 = (a_n + 1)^2$. Thus,

$$a_n + 1 = (a_1 + 1)^{2^{n-1}} = 8^{2^{n-1}} = 2^{3 \times 2^{n-1}},$$

i.e. $a_n = 2^{3 \times 2^{n-1}} - 1$.

It is obvious that $\{a_n\}$ is strictly increasing. Since

$$a_{11} = 2^{3072} - 1 < 2^{4036}, a_{12} = 2^{6144} - 1 > 2^{4036} = 4^{2018},$$

the smallest positive integer n that satisfies the given condition is 12. □

10 (20 marks) Given function $f(x) = \begin{cases} |\log_3 x - 1|, & 0 < x \leqslant 9, \\ 4 - \sqrt{x}, & x > 9. \end{cases}$

defined on $\mathbb{R}^+$, let a, b, c be three real numbers that are not equal to each other satisfying $f(a) = f(b) = f(c)$. Find the range of abc.

Solution   Without loss of generality, suppose $a < b < c$. Since $f(x)$ is strictly decreasing on $(0, 3]$, strictly increasing onor $[3, 9]$, and strictly decreasing on $[9, +\infty)$ with $f(3) = 0, f(9) = 1$, then from the graph of the function, we find

$$a \in (0, 3), b \in (3, 9), c \in (9, +\infty),$$

and also $f(a) = f(b) = f(c) \in (0, 1)$.

From $f(a) = f(b)$, we have

$$1 - \log_3 a = \log_3 b - 1,$$

or $\log_3 a + \log_3 b = 2$. Then $ab = 3^2 = 9$, and $abc = 9c$.

From
$$0 < f(c) = 4 - \sqrt{c} < 1,$$
we get $c \in (9, 16)$. Hence $abc = 9c \in (81, 144)$.

Therefore, the range of abc is $(81, 144)$. □

Remark For any $r \in (81, 144)$, take $c_0 = \dfrac{r}{9}$. Then $c_0 \in (9, 16)$, and hence $f(c_0) \in (0, 1)$. Draw line l passing through point $(c_0, f(c_0))$ and parallel to the x-axis. Then l intersects with the graph of $f(x)$ at two other points: $(a, f(a)), (b, f(b))$, where $a \in (0, 3), b \in (3, 9)$. $f(a) = f(b) = f(c)$ and $ab = 9$ are satisfied. Therefore, $abc = r$.

11 (20 marks) As shown in Fig. 11.1, in a plane rectangular coordinate system xOy, A, B, C, D are the left, right, top, and bottom vertices of ellipse $\Gamma : \dfrac{x^2}{a^2} + \dfrac{y^2}{b^2} = 1$ $(a > b > 0)$, respectively. Points P, Q are on Γ and in the first quadrant, satisfying $OQ // AP$. M is the midpoint of line segment AP, and ray OM intersects the ellipse at point R. Prove that line segments OQ, OR, BC form a right triangle.

Solution Suppose the coordinates of P are (x_0, y_0). Since $\overrightarrow{OQ} // \overrightarrow{AP}$ and $\overrightarrow{OR} // \overrightarrow{OM}$, where $\overrightarrow{AP} = \overrightarrow{OP} - \overrightarrow{OA}$ and $\overrightarrow{OM} = \dfrac{1}{2}(\overrightarrow{OP} + \overrightarrow{OA})$, there exist real numbers λ, μ such that
$$\overrightarrow{OQ} = \lambda(\overrightarrow{OP} - \overrightarrow{OA}), \overrightarrow{OR} = \mu(\overrightarrow{OP} + \overrightarrow{OA}).$$

Here, the coordinates of points Q, R can be expressed as $(\lambda(x_0 + a), \lambda y_0), (\mu(x_0 - a), \mu y_0)$, respectively. Since P, Q are on the ellipse,
$$\lambda^2 \left(\frac{(x_0 + a)^2}{a^2} + \frac{y_0^2}{b^2} \right) = \mu^2 \left(\frac{(x_0 - a)^2}{a^2} + \frac{y_0^2}{b^2} \right) = 1.$$

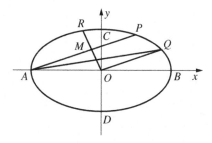

Fig. 11.1

Since $\dfrac{x_0^2}{a^2} + \dfrac{y_0^2}{b^2} = 1$, the equation is simplified to $\lambda^2 \left(2 + \dfrac{2x_0}{a}\right) = \mu^2 \left(2 - \dfrac{2x_0}{a}\right) = 1$, whose solution is

$$\lambda^2 = \frac{a}{2\,(a + x_0)}, \quad \mu^2 = \frac{a}{2\,(a - x_0)}.$$

Therefore,

$$
\begin{aligned}
|OQ|^2 + |OR|^2 &= \lambda^2((x_0 + a)^2 + y_0^2) + \mu^2((x_0 - a)^2 + y_0^2) \\
&= \frac{a}{2(a + x_0)}((x_0 + a)^2 + y_0^2) + \frac{a}{2(a - x_0)}((x_0 - a)^2 + y_0^2) \\
&= \frac{a(a + x_0)}{2} + \frac{a y_0^2}{2(a + x_0)} + \frac{a(a - x_0)}{2} + \frac{a y_0^2}{2(a - x_0)} \\
&= a^2 + \frac{a y_0^2}{2}\left(\frac{1}{a + x_0} + \frac{1}{a - x_0}\right) = a^2 + \frac{a y_0^2}{2} \cdot \frac{2a}{a^2 - x_0^2} \\
&= a^2 + \frac{a^2 \cdot b^2 \left(1 - \dfrac{x_0^2}{a^2}\right)}{a^2 - x_0^2} = a^2 + b^2 = |BC|^2.
\end{aligned}
$$

Consequently, line segments OQ, OR and BC form a right triangle. $\square$

China Mathematical Competition

2019

While the scope of the test questions in the first round of the 2019 China Mathematical Competition does not exceed the teaching requirements and content specified in the "General High School Mathematics Curriculum Standards (Experiments)" promulgated by the Ministry of Education of China in 2017, the methods of proposing the questions have been improved. The emphasis placed on to test the students' basic knowledge and skills, and their abilities to integrate and use flexibly of them. Each test paper includes eight fill-in-the-blank questions and three answer questions. The answer time is 80 minutes, and the full score is 120 points.

The scope of the test questions in the second round (Complementary Test) is in line with the International Mathematical Olympiad, with some expanded knowledge, plus a few contents of the Mathematical Competition Syllabus. Each test paper consists of four answer questions, including a plane geometry one, and the answering time is 170 minutes. The full score is 180 points.

Test Paper A
(8:00 – 9:20; September 8, 2019)

Part I Short-Answer Questions (Questions 1–8, eight marks each)

1 Suppose positive real number a satisfies $a^a = (9a)^{8a}$. Then the value of $\log_a(3a)$ is _____.

Solution   By the given condition, we have $9a = a^{\frac{1}{8}}$. Then

$$3a = \sqrt{9a \cdot a} = a^{\frac{9}{16}}.$$

Consequently,

$$\log_a(3a) = \frac{9}{16}. \qquad \square$$

2 Given real number set $\{1, 2, 3, x\}$, suppose the difference between its maximum element and minimum element is equal to the sum of its all elements. Then the value of x is _____.

Solution   If $x \geq 0$, then the difference between its maximum element and minimum element cannot exceed $\max\{3, x\}$. However, the sum of its all elements is greater than $\max\{3, x\}$, not agree with the given condition.

Therefore $x < 0$, namely, x is the minimum element. Then,

$$3 - x = 6 + x,$$

and this yields the solution $x = -\dfrac{3}{2}$. $\qquad \square$

3 In a plane rectangular coordinate system, $\overrightarrow{e}$ is a unit vector, vector $\overrightarrow{a}$ satisfies $\overrightarrow{a} \cdot \overrightarrow{e} = 2$, and $|\overrightarrow{a}|^2 \leq 5|\overrightarrow{a} + t\overrightarrow{e}|$ holds for any real number t. Then the range of $|\overrightarrow{a}|$ is _____.

Solution   Without loss of generality, let $\overrightarrow{e} = (1, 0)$. Since $\overrightarrow{a} \cdot \overrightarrow{e} = 2$, we can put $\overrightarrow{a} = (2, s)$. Then for any real number t, we have

$$4 + s^2 = |\overrightarrow{a}|^2 \leq 5|\overrightarrow{a} + t\overrightarrow{e}| = 5\sqrt{(2+t)^2 + s^2}.$$

This is equivalent to $4 + s^2 \leq 5\,|s|$. So the solution is $|s| \in [1, 4]$, namely, $s^2 \in [1, 16]$.

Hence, we obtain

$$|\vec{a}| = \sqrt{4 + s^2} \in [\sqrt{5}, 2\sqrt{5}].$$
□

4 Suppose A, B are the vertices of the major axis of ellipse Γ, E, F are the foci of Γ. It is given that $|AB| = 4, |AF| = 2 + \sqrt{3}$, and P is a point on Γ satisfying $|PE| \cdot |PF| = 2$. Then the area of $\triangle PEF$ is _____.

Solution In a plane rectangular coordinate system, without loss of generality, we assume the standard equation of ellipse Γ is

$$\frac{x^2}{a^2} + \frac{y^2}{b^2} = 1 \ (a > b > 0).$$

According to the given conditions and the geometrical properties of ellipse, it is easy to get

$$2a = |AB| = 4, \ a + \sqrt{a^2 - b^2} = |AF| = 2 + \sqrt{3}.$$

Then we have $a = 2, b = 1$, and $|EF| = 2\sqrt{a^2 - b^2} = 2\sqrt{3}$.

By definition of ellipse, it is known that

$$|PE| + |PF| = 2a = 4.$$

Combing with the given condition $|PE| \cdot |PF| = 2$, then

$$|PE|^2 + |PF|^2 = (|PE| + |PF|)^2 - 2\,|PE| \cdot |PF| = 12 = |EF|^2,$$

i.e., $\angle EPF$ is a right angle. Therefore,

$$S_{\triangle PEF} = \frac{1}{2} \cdot |PE| \cdot |PF| = 1.$$
□

5 Suppose a is randomly selected from $1, 2, 3, \ldots, 10$, and b is randomly selected from $-1, -2, -3, \ldots, -10$. Then the probability of $a^2 + b$ divisible by 3 is _____.

Solution There are $10^2 = 100$ equally likely selections for array (a, b). Let us consider the number of selections, say N, that $a^2 + b$ is divisible by 3.

If a is divisible by 3, then b is also divisible by 3. In this case, there are 3 selections for a, b, respectively. Then the number of selections for (a, b) is $3^2 = 9$.

If a is not divisible by 3, then $a^2 \equiv 1 \,(\text{mod } 3)$, and hence $b \equiv -1 \,(\text{mod } 3)$. In this case, a has 7 selections, and b has 4. Therefore (a, b) has $7 \times 4 = 28$ outcomes.

Hence $N = 9 + 28 = 37$, and the required probability is $\dfrac{37}{100}$. $\square$

6 For any closed interval I, we denote the maximum of function $y = \sin x$ on I by M_I. Suppose positive number a satisfies $M_{[0,a]} = 2M_{[a,2a]}$. Then the value of a is _____.

Solution Assuming $0 < a \leq \dfrac{\pi}{2}$, by the geometrical properties of the sine function, it follows that

$$0 < M_{[0,a]} = \sin a \leq M_{[a,2a]},$$

and this is inconsistent with the given condition.

Hence $a > \dfrac{\pi}{2}$, and then $M_{[0,a]} = 1$, therefore $M_{[a,2a]} = \dfrac{1}{2}$. Then there exists nonnegative integer k satisfying

$$2k\pi + \frac{5}{6}\pi \leq a < 2a \leq 2k\pi + \frac{13}{6}\pi, \qquad \textcircled{1}$$

Furthermore, at least one of "$\leq$" in Equation $\textcircled{1}$ gets the equal sign.

When $k = 0$, we get $a = \dfrac{5}{6}\pi$ or $2a = \dfrac{13}{6}\pi$.

By checking, $a = \dfrac{5}{6}\pi$ and $a = \dfrac{13}{12}\pi$ are both consistent with the given condition.

When $k \geq 1$, since $2k\pi + \dfrac{13}{6}\pi < 2\left(2k\pi + \dfrac{5}{6}\pi\right)$, there does not exist a satisfying Equation $\textcircled{1}$.

To sum up, the value of a is $\dfrac{5}{6}\pi$ or $\dfrac{13}{12}\pi$. $\square$

7 As shown in Fig. 7.1, a cross section of cube $ABCD - EFGH$ passes through vertices A, C and point K on edge EF, dividing the cube into two parts with a volume ratio of 3:1. Then the value of $\dfrac{EK}{KF}$ is _____.

Solution As shown in Fig. 7.2, take α as the plane of the cross section. Extend AK, BF to let them intersect at point P. Then P is on α, and

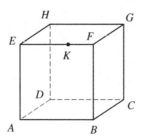

Fig. 7.1

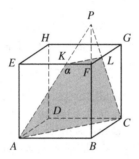

Fig. 7.2

line CP is the intersection of α and plane $BCGF$. Assume CP and FG intersect at point L, then quadrilateral $AKLC$ is the cross section.

Since plane ABC is parallel to plane KFL and lines AK, BF, CL intersect at a common point P, therefore $ABC - KFL$ is a frustum. Without loss of generality, suppose the edge length of the cube is 1, then its volume is 1, and, by the given condition, the volume of frustum $ABC - KFL$ is $V = \dfrac{1}{4}$.

Suppose $PF = h$. Then $\dfrac{KF}{AB} = \dfrac{FL}{BC} = \dfrac{PF}{PB} = \dfrac{h}{h+1}$. It should be noted that PB, PF are the heights of pyramids $P - ABC$ and $P - KFL$, respectively. Therefore,

$$\frac{1}{4} = V = V_{P-ABC} - V_{P-KFL}$$

$$= \frac{1}{6}AB \cdot BC \cdot PB - \frac{1}{6}KF \cdot FL \cdot PF$$

$$= \frac{1}{6}(h+1)\left(1 - \left(\frac{h}{h+1}\right)^3\right)$$

$$= \frac{3h^2 + 3h + 1}{6(h+1)^2}.$$

Through simplification, we find $3h^2 = 1$, and then $h = \dfrac{1}{\sqrt{3}}$.

Consequently,

$$\frac{EK}{KF} = \frac{AE}{PF} = \frac{1}{h} = \sqrt{3}. \qquad \square$$

8 Arranging six numbers $2, 0, 1, 9, 20, 19$ in any order to get an 8-digit number with the first digit not being 0. Then the number of different 8-digit numbers generated is _____.

Solution Let A be the set of all the permutations of $2, 0, 1, 9, 20, 19$ with the first digit not being zero. It is easy to find $|A| = 5 \times 5! = 600$ ($|X|$ stands for the number of elements of finite set X).

Denote by B the subset of all the permutations of A with the following term of 2 not being 0 and the following term of 1 not being 9. Denote by C the subset of all the permutations of A with the following term of 2 being 0 while the following term of 1 not being 9. Denote by D the subset of all the permutations of A with the following term of 1 being 9 while the following term of 2 not being 0.

It is easy to find that $|B| = 4!$, $|B| + |C| = 5!$, $|B| + |D| = 4 \times 4!$. Then

$$|B| = 24, |C| = 96, |D| = 72.$$

For each 8-digit number generated in B, it just corresponds to $2 \times 2 = 4$ permutations in B (In such permutations, 20 is interchangeable with "$2, 0$", and 19 is commutable with "$1, 9$"). Similarly, for each 8-digit number generated in C or D, it just corresponds to 2 permutations in C or D. Therefore, the number of the 8-digit numbers satisfying the given condition is

$$|A\backslash(B \cup C \cup D)| + \frac{|B|}{4} + \frac{|C| + |D|}{2}$$

$$= |A| - \frac{3|B|}{4} - \frac{|C|}{2} - \frac{|D|}{2}$$

$$= 600 - 18 - 48 - 36 = 498. \qquad \square$$

Part II Word Problems (16 marks for Question 9, 20 marks each for Question 10 and 11, and then 56 marks in total)

9 (16 marks) In $\triangle ABC, BC = a, CA = b, AB = c$. Suppose b is the geometric mean of a and c; $\sin A$ is the arithmetic mean of $\sin(B - A)$ and $\sin C$. Find the value of $\cos B$.

Solution Since b is the geometric mean of a and c, there exists $q > 0$ satisfying

$$b = qa, \quad c = q^2 a. \qquad \qquad (1)$$

Since $\sin A$ is the arithmetic mean of $\sin(B - A)$ and $\sin C$, it follows that

$$
\begin{aligned}
2 \sin A &= \sin(B - A) + \sin C \\
&= \sin(B - A) + \sin(B + A) \\
&= 2 \sin B \cos A
\end{aligned}
$$

According to the laws of sines and cosines, one can find

$$\frac{a}{b} = \frac{\sin A}{\sin B} = \cos A = \frac{b^2 + c^2 - a^2}{2bc},$$

namely, $b^2 + c^2 - a^2 = 2ac$.

Putting (1) into the above expression and simplifying it, we obtain

$$q^2 + q^4 - 1 = 2q^2, \quad \text{that is} \quad q^4 = q^2 + 1.$$

Therefore

$$q^2 = \frac{\sqrt{5} + 1}{2},$$

and hence

$$\cos B = \frac{c^2 + a^2 - b^2}{2ac} = \frac{q^4 + 1 - q^2}{2q^2} = \frac{1}{q^2} = \frac{\sqrt{5} - 1}{2}. \qquad \square$$

10 (20 marks) In a plane rectangular coordinate system xOy, circle Ω and parabola $\Gamma : y^2 = 4x$ have just one common point, and Ω is tangent to the x-axis at the focus F of Γ. Find the radius of Ω.

Solution It is easy to find that the coordinates of focus F of Γ are $(1, 0)$. Assume the radius of circle Ω is $r(r > 0)$. By symmetry, without loss of generality, suppose Ω is tangent to the above part of x-axis at point F. The equation of Ω is then

$$(x-1)^2 + (y-r)^2 = r^2. \tag{1}$$

Putting $x = \dfrac{y^2}{4}$ into Equation ① and simplifying the result, we have

$$\left(\frac{y^2}{4} - 1\right)^2 + y^2 - 2ry = 0.$$

It is obvious that $y > 0$, therefore

$$r = \frac{1}{2y}\left(\left(\frac{y^2}{4} - 1\right)^2 + y^2\right) = \frac{(y^2+4)^2}{32y}. \tag{2}$$

By the given condition, Equation ② has just one positive solution y which corresponds to the unique common point between Ω and Γ.

We shall consider the minimum value of $f(y) = \dfrac{(y^2+4)^2}{32y}$ $(y > 0)$.

Making use of the inequality of arithmetic and geometric means, we have

$$y^2 + 4 = y^2 + \frac{4}{3} + \frac{4}{3} + \frac{4}{3} \geq 4\sqrt[4]{y^2 \cdot \left(\frac{4}{3}\right)^3}.$$

Therefore

$$f(y) \geq \frac{1}{32y} \cdot 16\sqrt{y^2 \cdot \left(\frac{4}{3}\right)^3} = \frac{4\sqrt{3}}{9}.$$

So $f(y)$ takes the minimum value $\dfrac{4\sqrt{3}}{9}$ if and only if $y^2 = \dfrac{4}{3}$, namely $y = \dfrac{2\sqrt{3}}{3}$.

Substituting the above result into Equation ②, we get $r \geq \dfrac{4\sqrt{3}}{9}$.

Assuming $r > \dfrac{4\sqrt{3}}{9}$, since $f(y)$ changes continuously with y and becomes arbitrarily large when $y \to 0^+$ or $y \to +\infty$, therefore ② has solutions both in $\left(0, \dfrac{2\sqrt{3}}{3}\right)$ and $\left(\dfrac{2\sqrt{3}}{3}, +\infty\right)$. This leads to a contradiction to the uniqueness of solution.

In conclusion, only $r = \dfrac{4\sqrt{3}}{9}$ is consistent with the given condition (at this time $\left(\dfrac{1}{3}, \dfrac{2\sqrt{3}}{3}\right)$ is the unique common point between Ω and Γ). $\quad\square$

11 (20 marks) A complex number sequence $\{z_n\}$ is called "funny" if $|z_1| = 1$ and $4z_{n+1}^2 + 2z_n z_{n+1} + z_n^2 = 0$ holds for any positive integer n. Find the maximum constant C such that $|z_1 + z_2 + \cdots + z_m| \geq C$ holds for any funny sequence $\{z_n\}$ and any positive integer m.

Solution Consider a funny complex number sequence $\{z_n\}$. It is known by induction that $z_n \neq 0$ $(n \in \mathbb{N}^+)$. By the given condition, we have

$$4\left(\frac{z_{n+1}}{z_n}\right)^2 + 2\left(\frac{z_{n+1}}{z_n}\right) + 1 = 0 \ (n \in \mathbb{N}^+).$$

The solution is

$$\frac{z_{n+1}}{z_n} = \frac{-1 \pm \sqrt{3}\,\mathrm{i}}{4} \ (n \in \mathbb{N}^+).$$

Furthermore,

$$\frac{|z_{n+1}|}{|z_n|} = \left|\frac{z_{n+1}}{z_n}\right| = \left|\frac{-1 + \sqrt{3}\,\mathrm{i}}{4}\right| = \frac{1}{2}.$$

Consequently,

$$|z_n| = |z_1| \cdot \frac{1}{2^{n-1}} = \frac{1}{2^{n-1}} \ (n \in \mathbb{N}^+). \qquad \text{①}$$

And then

$$|z_n + z_{n+1}| = |z_n| \cdot \left|1 + \frac{z_{n+1}}{z_n}\right| = \frac{1}{2^{n-1}} \cdot \left|\frac{3 \pm \sqrt{3}\,\mathrm{i}}{4}\right| = \frac{\sqrt{3}}{2^n} \ (n \in \mathbb{N}^+). \ \text{②}$$

Take $T_m = |z_1 + z_2 + \cdots + z_m| \ (m \in \mathbb{N}^+)$.

When $m = 2s$ $(s \in \mathbb{N}^+)$, by Equation ② we find

$$T_m \geq |z_1 + z_2| - \sum_{k=2}^{s} |z_{2k-1} + z_{2k}|$$

$$> \frac{\sqrt{3}}{2} - \sum_{k=2}^{\infty} |z_{2k-1} + z_{2k}|$$

$$= \frac{\sqrt{3}}{2} - \sum_{k=2}^{\infty} \frac{\sqrt{3}}{2^{2k-1}} = \frac{\sqrt{3}}{3}.$$

When $m = 2s + 1$ $(s \in \mathbb{N}^+)$, from ① and ② we have

$$|z_{2s+1}| = \frac{1}{2^{2s}} < \frac{\sqrt{3}}{3 \cdot 2^{2s-1}} = \sum_{k=s+1}^{\infty} \frac{\sqrt{3}}{2^{2k-1}} = \sum_{k=s+1}^{\infty} |z_{2k-1} + z_{2k}|.$$

Hence

$$T_m \geq |z_1 + z_2| - \left(\sum_{k=2}^{s} |z_{2k-1} + z_{2k}| \right) - |z_{2s+1}|$$

$$> \frac{\sqrt{3}}{2} - \sum_{k=2}^{\infty} |z_{2k-1} + z_{2k}|$$

$$= \frac{\sqrt{3}}{3}.$$

When $m = 1$, $T_1 = |z_1| = 1 > \frac{\sqrt{3}}{3}$.

The above discussion indicates that $C = \frac{\sqrt{3}}{3}$ satisfies the given condition.

On the other hand, when $z_1 = 1$, $z_{2k} = \frac{-1 + \sqrt{3}\,\mathrm{i}}{2^{2k}}$, $z_{2k+1} = \frac{-1 - \sqrt{3}\,\mathrm{i}}{2^{2k+1}}$ $(k \in \mathbb{N}^+)$, it is easy to verify that $\{z_n\}$ is a funny sequence. Then

$$\lim_{s \to \infty} T_{2s+1} = \lim_{s \to \infty} \left| z_1 + \sum_{k=1}^{s} (z_{2k} + z_{2k+1}) \right|$$

$$= \lim_{s \to \infty} \left| 1 + \sum_{k=1}^{s} \frac{-3 + \sqrt{3}\,\mathrm{i}}{2^{2k+1}} \right|$$

$$= \left| 1 + \frac{-3 + \sqrt{3}\,i}{8} \cdot \frac{4}{3} \right|$$

$$= \frac{\sqrt{3}}{3}.$$

This indicates that C cannot exceed $\dfrac{\sqrt{3}}{3}$.

In conclusion, the desired solution of C is $\dfrac{\sqrt{3}}{3}$. □

Test Paper B
(8:00 – 9:20; September 8, 2019)

Part I Short-Answer Questions (Questions 1–8, eight marks each)

1 Given real number set $\{1, 2, 3, x\}$, its maximum element is equal to the sum of its all elements. Then the value of x is _____.

Solution The given condition is equivalent to the statement that the sum of the other elements besides the maximum element is 0.

It is immediately obvious that $x < 0$. Then

$$1 + 2 + x = 0.$$

Therefore, $x = -3$. □

2 Suppose vector $\overrightarrow{a} = (2^m, -1)$ is perpendicular to vector $\overrightarrow{b} = (2^m - 1, 2^m + 1)$, where m is a real number. Then the magnitude of $\overrightarrow{a}$ is _____.

Solution Let $2^m = t$, $t > 0$. The given condition can be equivalently expressed as

$$t \cdot (t - 1) + (-1) \cdot 2t = 0.$$

The solution is $t = 3$. Therefore, the magnitude of $\overrightarrow{a}$ is $\sqrt{3^2 + (-1)^2} = \sqrt{10}$. □

3 Suppose $\alpha, \beta \in (0, \pi)$, $\cos\alpha$ and $\cos\beta$ are the two roots of equation $5x^2 - 3x + 1 = 0$. Then the value of $\sin\alpha \sin\beta$ is _____.

Solution   By the given condition, we have $\cos\alpha + \cos\beta = \dfrac{3}{5}$, $\cos\alpha\cos\beta = -\dfrac{1}{5}$. Therefore,

$$
\begin{aligned}
(\sin\alpha\sin\beta)^2 &= (1 - \cos^2\alpha)(1 - \cos^2\beta) \\
&= 1 - \cos^2\alpha - \cos^2\beta + \cos^2\alpha\cos^2\beta \\
&= (1 + \cos\alpha\cos\beta)^2 - (\cos\alpha + \cos\beta)^2 \\
&= \left(\frac{4}{5}\right)^2 - \left(\frac{3}{5}\right)^2 = \frac{7}{25}.
\end{aligned}
$$

Since $\alpha, \beta \in (0, \pi)$, it follows that $\sin\alpha\sin\beta > 0$. Thus $\sin\alpha\sin\beta = \dfrac{\sqrt{7}}{5}$. □

4 Suppose triangular pyramid $P - ABC$ satisfies $PA = PB = 3$, $AB = BC = CA = 2$. Then the maximum volume of the triangular pyramid is _____.

Solution   We denote the height of triangular pyramid $P - ABC$ by h, and take M as the midpoint of edge AB, then

$$
h \le PM = \sqrt{3^2 - 1^2} = 2\sqrt{2}.
$$

When plane PAB is perpendicular to plane ABC, h has the largest value $2\sqrt{2}$. Then the volume of the triangular pyramid takes its maximum

$$
\frac{1}{3}S_{\triangle ABC} \cdot 2\sqrt{2} = \frac{1}{3} \cdot \sqrt{3} \cdot 2\sqrt{2} = \frac{2\sqrt{6}}{3}. \qquad \square
$$

5 Arranging five numbers 2, 0, 1, 9, 2019 on a line in any order to generate an 8-digit number (with the first digit not being zero). Then the number of the different 8-digit numbers generated is _____.

Solution   It is easy to observe that the permutations of 2, 0, 1, 9, 2019 with the first digit not being 0 is $4 \times 4! = 96$.

However, permutations $(2, 0, 1, 9, 2019)$ and $(2019, 2, 0, 1, 9)$ correspond to the same number 20 192 019, while the other numbers generated are different from each other. So the number of 8-digit number satisfying the given condition is $96 - 1 = 95$. □

6 Given integer $n > 4$, in the expansion of $(x + 2\sqrt{y} - 1)^n$, the coefficients of terms x^{n-4} and xy are the same. Then the value of n is

_____.

Solution Note that $(x + 2\sqrt{y} - 1)^n = \sum_{r=0}^{n} C_n^r x^{n-r} (2\sqrt{y} - 1)^r$. Term x^{n-4} appears only in the expansion $C_n^4 x^{n-4} (2\sqrt{y}-1)^4$, so the corresponding coefficient is

$$(-1)^4 C_n^4 = \frac{n(n-1)(n-2)(n-3)}{24}.$$

On the other hand, term xy appears only in the expansion $C_n^{n-1} x (2\sqrt{y} - 1)^{n-1}$, so the corresponding coefficient is

$$C_n^{n-1} C_{n-1}^2 4 \cdot (-1)^{n-3} = (-1)^{n-3} 2n(n-1)(n-2).$$

Thus

$$\frac{n(n-1)(n-2)(n-3)}{24} = (-1)^{n-3} 2n(n-1)(n-2).$$

Note that $n > 4$. Simplifying the above equation leads to $n - 3 = (-1)^{n-3} 48$.

Therefore, n can only be an odd number and $n - 3 = 48$. The solution is $n = 51$. $\square$

7 In a plane rectangular coordinate system, suppose there exists a point (a, b) on the circle with center coordinates $(r + 1, 0)$ and radius r satisfying $b^2 \geq 4a$. Then the minimum value of r is _____.

Solution By the given condition we have $(a - r - 1)^2 + b^2 = r^2$. Thus

$$4a \leq b^2 = r^2 - (a - r - 1)^2 = 2r(a - 1) - (a - 1)^2.$$

That is

$$a^2 - 2(r - 1)a + 2r + 1 \leq 0.$$

The above quadratic inequality of a has solutions, therefore the discriminant

$$[2(r - 1)]^2 - 4(2r + 1) = 4r(r - 4) \geq 0,$$

The solution is $r \geq 4$.

After checking, $(a, b) = (3, 2\sqrt{3})$ satisfies the given condition when $r \geq 4$. Therefore, the minimum value of r is 4. $\square$

8 Suppose all the terms of arithmetic sequence $\{a_n\}$ are integers with the initial term $a_1 = 2019$, and for any positive integer n, there always exists positive integer m such that $a_1 + a_2 + \cdots + a_n = a_m$. Then the number of such sequences is _____.

Solution   Let the common difference of $\{a_n\}$ be d. By the given condition we know that $a_1 + a_2 = a_k$ (k is some positive integer). Then

$$2a_1 + d = a_1 + (k-1)d,$$

namely $(k-2)d = a_1$. We have $k \neq 2$ and $d = \dfrac{a_1}{k-2}$. Hence

$$a_n = a_1 + (n-1)d = a_1 + \frac{n-1}{k-2}a_1,$$

For any positive integer n, since

$$a_1 + a_2 + \cdots + a_n = a_1 n + \frac{n(n-1)}{2}d$$

$$= a_1 + (n-1)a_1 + \frac{n(n-1)}{2}d$$

$$= a_1 + \left((n-1)(k-2) + \frac{n(n-1)}{2} \right)d,$$

the above is really a term in $\{a_n\}$.

Therefore, we only need to consider the number of positive integers k such that $(k-2) \mid a_1$ is valid. Note that 2019 is the product of primes 3 and 673, then it is easy to find that the values of $k-2$ are $-1, 1, 3, 673, 2019$, which correspond to 5 arithmetic sequences satisfying the given condition. $\square$

Part II Word Problems (16 marks for Question 9, 20 marks each for Question 10 and 11, and then 56 marks in total)

 9   (16 marks) In ellipse Γ, F is one of the foci, A, B are two vertices. If $|FA| = 3$, $|FB| = 2$, then find all the possible values of $|AB|$.

Solution   In a plane rectangular coordinate system, without loss of generality, we assume the standard equation of ellipse Γ is

$$\frac{x^2}{a^2} + \frac{y^2}{b^2} = 1 \, (a > b > 0),$$

and denote $c = \sqrt{a^2 - b^2}$. By symmetry, suppose F is the right focus of Γ.

It is easy to find that the distance from F to the left vertex of Γ is $a+c$, that to the left vertex is $a-c$, and that to the top and bottom vertices are both a. The discussion shall be made in the following situations.

(1) Suppose A, B are the left and right vertices, respectively. Then $a + c = 3$, $a - c = 2$, and $|AB| = 2a = 5$. Accordingly, $b^2 = (a+c)(a-c) = 6$, so the equation of ellipse Γ is $\dfrac{4x^2}{25} + \dfrac{y^2}{6} = 1$.

(2) Suppose A is the left vertex and B is the top or bottom vertex. Then $a + c = 3$, $a = 2$, and $c = 1$. Hence

$$b^2 = a^2 - c^2 = 3,$$

so $|AB| = \sqrt{a^2 + b^2} = \sqrt{7}$. The corresponding equation of ellipse Γ is $\dfrac{x^2}{4} + \dfrac{y^2}{3} = 1$.

(3) Suppose A is the top or bottom vertex and B is the right vertex. Then $a = 3$, $a - c = 2$, and $c = 1$. Hence

$$b^2 = a^2 - c^2 = 8,$$

so $|AB| = \sqrt{a^2 + b^2} = \sqrt{17}$. The corresponding equation of the ellipse Γ is $\dfrac{x^2}{9} + \dfrac{y^2}{8} = 1$.

To sum up, all the possible values of $|AB|$ are $5, \sqrt{7}, \sqrt{17}$. □

10 (20 marks) Suppose a, b, c all are greater than 1 and satisfy

$$\begin{cases} \lg a + \log_b c = 3, \\ \lg b + \log_a c = 4. \end{cases}$$

Find the maximum value of $\lg a \cdot \lg c$.

Solution Let $\lg a = x$, $\lg b = y$, $\lg c = z$. Since $a, b, c > 1$, we have $x, y, z > 0$.

By the given condition and the change-of-base formula, we find

$$x + \frac{z}{y} = 3, \quad y + \frac{z}{x} = 4.$$

That is

$$xy + z = 3y = 4x.$$

Let $x = 3t$, $y = 4t$ $(t > 0)$. Then

$$z = 4x - xy = 12t - 12t^2.$$

And from $z > 0$ we can observe that $t \in (0, 1)$.

Therefore, making use of the generalized mean inequality for three numbers, we have

$$
\begin{aligned}
\lg a \lg c = xz &= 3t \cdot 12t(1 - t) \\
&= 18 \cdot t^2(t - 2t) \\
&\leq 18 \cdot \left(\frac{t + t + (2 - 2t)}{3} \right)^3 \\
&= 18 \cdot \left(\frac{2}{3} \right)^3 = \frac{16}{3}.
\end{aligned}
$$

When $t = 2 - 2t$, namely $t = \dfrac{2}{3}$, the corresponding a, b, c are $100, 10^{\frac{8}{3}}$, $10^{\frac{8}{3}}$, respectively. Then $\lg a \lg c$ takes its maximum value $\dfrac{16}{3}$. $\qquad\square$

11 (20 marks) Suppose complex sequence $\{z_n\}$ satisfies $|z_1| = 1$, and $4z_{n+1}^2 + 2z_n z_{n+1} + z_n^2 = 0$ for any positive integer n.

Prove that $|z_1 + z_2 + \cdots + z_m| \leq \dfrac{2\sqrt{3}}{3}$ for any positive integer m.

Proof It is known by induction that $z_n \neq 0 \ (n \in \mathbb{N}^+)$. By the given condition, we have

$$
4 \left(\frac{z_{n+1}}{z_n} \right)^2 + 2 \left(\frac{z_{n+1}}{z_n} \right) + 1 = 0 \ (n \in \mathbb{N}^+),
$$

The solution is

$$
\frac{z_{n+1}}{z_n} = \frac{-1 \pm \sqrt{3}\,\mathrm{i}}{4} \ (n \in \mathbb{N}^+).
$$

Furthermore,

$$
\frac{|z_{n+1}|}{|z_n|} = \left| \frac{z_{n+1}}{z_n} \right| = \left| \frac{-1 + \sqrt{3}\,\mathrm{i}}{4} \right| = \frac{1}{2}.
$$

Therefore,

$$
|z_n| = |z_1| \cdot \frac{1}{2^{n-1}} = \frac{1}{2^{n-1}} \ (n \in \mathbb{N}^+). \tag{1}
$$

And then

$$
|z_n + z_{n+1}| = |z_n| \cdot \left| 1 + \frac{z_{n+1}}{z_n} \right| = \frac{1}{2^{n-1}} \cdot \left| \frac{3 \pm \sqrt{3}\mathrm{i}}{4} \right| = \frac{\sqrt{3}}{2^n} \ (n \in \mathbb{N}^+). \tag{2}
$$

If m is an even number, let $m = 2s$ ($s \in \mathbb{N}^+$), by making use of $\textcircled{2}$ we find

$$|z_1 + z_2 + \cdots + z_m| \le \sum_{k=1}^{s} |z_{2k-1} + z_{2k}|$$

$$< \sum_{k=1}^{\infty} |z_{2k-1} + z_{2k}|$$

$$= \sum_{k=1}^{\infty} \frac{\sqrt{3}}{2^{2k-1}} = \frac{2\sqrt{3}}{3}.$$

If m is an odd number, let $m = 2s + 1$ ($s \in \mathbb{N}^+$), from $\textcircled{1}$, $\textcircled{2}$ we have

$$|z_{2s+1}| = \frac{1}{2^{2s}} < \frac{\sqrt{3}}{3 \cdot 2^{2s-1}} = \sum_{k=s+1}^{\infty} \frac{\sqrt{3}}{2^{2k-1}} = \sum_{k=s+1}^{\infty} |z_{2k-1} + z_{2k}|.$$

Hence

$$|z_1 + z_2 + \cdots + z_m| \le \left(\sum_{k=1}^{s} |z_{2k-1} + z_{2k}| \right) + |z_{2s+1}|$$

$$< \sum_{k=1}^{\infty} |z_{2k-1} + z_{2k}| = \frac{2\sqrt{3}}{3}.$$

In conclusion, the proof is completed. $\qquad\square$

China Mathematical Competition (Complementary Test)

2018

Test Paper A
(9:40 – 12:10; September 9, 2018)

1 (40 marks) Let n be a positive integer, and $a_1, a_2, \ldots,$ $a_n, b_1, b_2, \ldots, b_n, A, B$ be positive real numbers, satisfying $a_i \leq b_i$, $a_i \leq A$, $i = 1, 2, \ldots, n$, $\dfrac{b_1 b_2 \ldots b_n}{a_1 a_2 \ldots a_n} \leq \dfrac{B}{A}$.

Prove that

$$\frac{(b_1 + 1)(b_2 + 1) \ldots (b_n + 1)}{(a_1 + 1)(a_2 + 1) \cdots (a_n + 1)} \leq \frac{B}{A}.$$

Solution By the given condition, we get $k_i = \dfrac{b_i}{a_i} \geq 1$, $i = 1, 2, \ldots, n$.

Denote $\dfrac{B}{A} = K$, then $\dfrac{b_1 b_2 \ldots b_n}{a_1 a_2 \ldots a_n} \leq \dfrac{B}{A}$ can be simplified as $k_1 k_2 \ldots k_n \leq K$.

Then we need to prove

$$\prod_{i=1}^{n} \frac{k_i a_i + 1}{a_i + 1} \leq \frac{KA + 1}{A + 1}. \qquad \qquad ①$$

For $i = 1, 2, \ldots, n$, since $k_i \geq 1$ and $0 < a_i \leq A$, it follows that

$$\frac{k_i a_i + 1}{a_i + 1} = k_i - \frac{k_i - 1}{a_i + 1} \leq k_i - \frac{k_i - 1}{A + 1} = \frac{k_i A + 1}{A + 1}.$$

Combing $K \geq k_1 k_2 \ldots k_n$, in order to prove ①, we only need to prove

$$\prod_{i=1}^{n} \frac{k_i A + 1}{A + 1} \leq \frac{k_1 k_2 \ldots k_n A + 1}{A + 1} \qquad \text{②}$$

for $A > 0, k_i \geq 1$ $(i = 1, 2, \ldots, n)$.

This can be proved by making induction on n. The conclusion clearly holds for $n = 1$.

When $n = 2$, since $A > 0, k_1, k_2 \geq 1$, we have

$$\frac{k_1 A + 1}{A + 1} \cdot \frac{k_2 A + 1}{A + 1} - \frac{k_1 k_2 A + 1}{A + 1} = -\frac{A(k_1 - 1)(k_2 - 1)}{(A + 1)^2} \leq 0. \qquad \text{③}$$

Therefore, the conclusion holds for $n = 2$.

Assume the conclusion is valid for $n = m$. Then when $n = m + 1$, by the induction hypothesis we find

$$\prod_{i=1}^{m+1} \frac{k_i A + 1}{A + 1} = \left(\prod_{i=1}^{m} \frac{k_i A + 1}{A + 1} \right) \cdot \frac{k_{m+1} A + 1}{A + 1}$$

$$\leq \frac{k_1 k_2 \ldots k_m A + 1}{A + 1} \cdot \frac{k_{m+1} A + 1}{A + 1}$$

$$\leq \frac{k_1 k_2 \ldots k_{m+1} A + 1}{A + 1}.$$

Finally, replace k_1, k_2 with $k_1 k_2 \ldots k_m, k_{m+1}$, respectively in ③ (Note that $k_1 k_2 \ldots k_m \geq 1$, $k_{m+1} \geq 1$), and hence the conclusion holds for $n = m + 1$.

By the mathematical induction method, ② holds for all positive integers n. Therefore, the proof is completed. $\qquad \square$

2 (40 marks) As shown in Fig. 2.1, in acute triangle $\triangle ABC$, $AB < AC$, M is the midpoint of side BC. Points D and E are the midpoints of the circumscribed circle arcs $\overparen{BAC}$ and $\overparen{BC}$ of $\triangle ABC$, respectively. F is the tangent point of the inscribed circle of $\triangle ABC$ on side AB, and G is the intersection of AE and BC. N lies on line EF satisfying $NB \perp AB$. Prove: if $BN = EM$, then $DF \perp FG$.

Solution By the given condition, DE is the diameter of the circumscribed circle of $\triangle ABC$, $DE \perp BC$ at M and $AE \perp AD$.

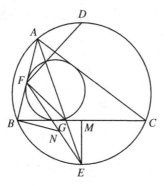

Fig. 2.1

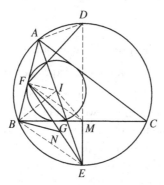

Fig. 2.2

As shown in Fig. 2.2, let I be the incenter of triangle $\triangle ABC$. Then I is on line AE and $IF \perp AB$.

Since $NB \perp AB$, then

$$\angle NBE = \angle ABE - \angle ABN$$

$$= (180° - \angle ADE) - 90° \qquad \text{(1)}$$

$$= 90° - \angle ADE = \angle MEI.$$

Also according to the property of the incenter, there is

$$\angle EBI = \angle EBC + \angle CBI = \angle EAC + \angle ABI$$

$$= \angle EAB + \angle ABI = \angle EIB,$$

and thus $BE = EI$.

Combing $BN = EM$ and ①, it follows that $\triangle NBE \cong MEI$.

Thus $\angle EMI = \angle BNE = 90° + \angle BFE = 180° - \angle EFI$, and hence points A, F, G, M are concyclic.

Furthermore, we find $\angle AFM = 90° + \angle IFM = 90° + \angle IEM = \angle AGM$. Therefore, points E, F, I, M are concyclic.

Then by $\angle DAG = \angle DMG = 90°$, it follows that points A, G, M, D are concyclic. Then A, F, G, M, D are also concyclic. So $\angle DFG = \angle DAG = 90°$, namely $DF \perp FG$. □

3 (50 marks) Let n, k, m be positive integers satisfying $k \geq 2$ and $n \leq m < \dfrac{2k-1}{k}n$. Suppose A is an n-element subset of $\{1, 2, \ldots, m\}$. Prove that every integer in $\left(0, \dfrac{n}{k-1}\right)$ can be expressed as $a - a'$, where $a, a' \in A$.

Solution　We shall use the proof by contradiction. Suppose there exists integer x in $\left(0, \dfrac{n}{k-1}\right)$ that cannot be expressed $a - a'$, where $a, a' \in A$. By the division with remainder, we get $m = xq + r$, $0 \leq r < x$. According the congruence class of modulo x, we divide $1, 2, \ldots, m$ into x arithmetic sequences with common difference x, where r arithmetic sequences have $q + 1$ terms and $x - r$ arithmetic sequences contain q terms. Since no two numbers in A have a difference of x, then A cannot contain two adjacent terms of an arithmetic sequence with common difference x. Thus we have

$$n = |A| \leq r\left\lceil \frac{q+1}{2} \right\rceil + (x-r)\left\lceil \frac{q}{2} \right\rceil = \begin{cases} x \cdot \dfrac{q+1}{2}, & 2 \nmid q, \\[2mm] x \cdot \dfrac{q}{2} + r, & 2 \mid q, \end{cases} \qquad ①$$

where $\lceil \alpha \rceil$ denotes the smallest integer not less than α.

By the given condition, we get

$$n > \frac{k}{2k-1}m = \frac{k}{2k-1}(xq + r). \qquad ②$$

Because $x \in \left(0, \dfrac{n}{k-1}\right)$, then

$$n > (k-1)x. \qquad ③$$

Case 1: q is odd. From ① we obtain

$$n \leq x \cdot \frac{q+1}{2}. \qquad ④$$

Combing ② and ④, it follows that $x \cdot \dfrac{q+1}{2} \geq n > \dfrac{k}{2k-1}(xq+r) \geq \dfrac{k}{2k-1}xq$, and hence $q < 2k-1$. Since q is odd, then $q \leq 2k+3$, and thus

$$n \leq x \cdot \frac{q+1}{2} \leq (k-1)x,$$

in contradiction to ③.

Case 2: q is odd. From ① we find

$$n \leq x \cdot \frac{q}{2} + r. \tag{5}$$

Combing ② and ⑤ leads to $x \cdot \dfrac{q}{2} + r \geq n > \dfrac{k}{2k-1}(xq+r)$, then

$$\frac{xq}{2(2k-1)} < \frac{k-1}{2k-1}r < \frac{(k-1)x}{2k-1},$$

and thus $q < 2k-1$. Since q is even, then $q \leq 2k-4$, and thus

$$n \leq x \cdot \frac{q}{2} + r \leq (k-2)x + r < (k-1)x,$$

in contradiction to ③.

In summary, the contradiction hypothesis does not hold and the conclusion is established. □

4 (50 marks) Sequence $\{a_n\}$ is defined as follows: a_1 is any positive integer; for integers $n \geq 1$, a_{n+1} is coprime with $\sum_{i=1}^{n} a_i$, as well as the smallest positive integer not equal to any of $a_1, a_2, \ldots, a_n$.

Prove that every positive integer occurs in sequence $\{a_n\}$.

Solution It is obvious that either $a_1 = 1$ or $a_2 = 1$. In the following we consider integer $m > 1$, and assume m has k distinct prime factors. We will prove by induction on k that m occurs in $\{a_n\}$. Put $S_n = a_1 + \cdots + a_n$, $n \geq 1$.

When $k = 1$, m is a prime power. Let $m = p^\alpha$, where $\alpha > 0$ and p is a prime number. Assume that m does not occur in $\{a_n\}$. Since the terms in $\{a_n\}$ are not identical to each other, then there exist positive integer N, such that $a_n > p^\alpha$, for $n \geq N$. If for some $n \geq N$, $p \nmid S_n$, then p^α and S_n are coprime. Since no term in $\{a_n\}$ is p^α, then by the definition of the sequence, we have $a_{n+1} \leq p^\alpha$, but this is in contradiction to $a_{n+1} > p^\alpha$.

Therefore $p \mid S_n$, for every $n \geq N$. However, from $p \mid S_{n+1}$ and $p \mid S_n$ we have $p \mid a_{n+1}$. Thus a_{n+1} and S_n are not coprime, contradicting the definition of a_{n+1}.

Suppose that $k \geq 2$ and the assertion holds for $k - 1$. Let the standard decomposition of m be $m = p_1^{\alpha_1} p_2^{\alpha_2} \ldots p_k^{\alpha_k}$. Assume m does not occur in $\{a_n\}$. Then there exists positive integer N', such that $a_n > m$ when $n \geq N'$. Take sufficiently large positive integers $\beta_1, \beta_2, \ldots, \beta_{k-1}$ such that

$$M = p_1^{\beta_1} p_2^{\beta_2} \ldots p_{k-1}^{\beta_{k-1}} > \max_{1 \leq n \leq N'} a_n.$$

We will prove that $a_{n+1} \neq M$, for $n \geq N$.

For any $n \geq N'$, if S_n and $p_1 p_2 \ldots p_k$ are coprime, then m and S_n are coprime. m does not occur in $\{a_n\}$, but $a_{n+1} > m$, contradicting the definition of the sequence. Therefore, we have:

For any $n \geq N'$, S_n and $p_1 p_2 \ldots p_k$ are not coprime.   (*)

Case 1: There exists i $(1 \leq i \leq k - 1)$ such that $p_i \mid S_n$. Since $(a_{n+1}, S_n) = 1$, then $p_i \nmid a_{n+1}$, and hence $a_{n+1} \neq M$ (because $p_i \mid M$).

Case 2: $p_i \nmid S_n$, for every i $(1 \leq i \leq k - 1)$. Then from (*), we have $p_k \mid S_n$. Therefore, $p_k \nmid a_{n+1}$, and then $p_k \nmid S_n + a_{n+1}$, which means $p_k \nmid S_{n+1}$. Therefore, from (*), we know there exists i_0 $(1 \leq i_0 \leq k - 1)$ such that $p_{i_0} \mid S_{n+1}$. But $S_{n+1} = S_n + a_{n+1}$ and $p_i \nmid S_n$ $(1 \leq i \leq k - 1)$. It follows that $p_{i_0} \nmid a_{n+1}$, and hence $a_{n+1} \neq M$.

Therefore, for $n \geq N' + 1$, $a_n \neq M$. But $M > \max\limits_{1 \leq n \leq N'} a_n$, then M does not occur in $\{a_n\}$, which is in contradiction to the induction hypothesis for $n = k - 1$. Consequently, if m has k distinct prime factors, then m must occur in $\{a_n\}$.

By mathematical induction, we know that all the positive integers occur in $\{a_n\}$. $\square$

Test Paper B
(9:40 – 12:10; September 9, 2018)

1 (40 marks) Suppose a, b are real numbers and a function is

$$f(x) = ax + b + \frac{9}{x}.$$

Prove that there exists $x_0 \in [1, 9]$ such that $|f(x_0)| \geq 2$.

Solution 1 We only need to prove that there exist $u, v \in [1, 9]$ satisfying $|f(u) - f(v)| \geq 4$. Then we have

$$|f(u)| + |f(v)| \geq |f(u) - f(v)| \geq 4,$$

and thus at least one of $|f(u)| \geq 2$ and $|f(v)| \geq 2$ is valid.

When $a \in \left(-\infty, \dfrac{1}{2}\right] \cup \left[\dfrac{3}{2}, +\infty\right)$, we have

$$|f(1) - f(9)| = |(a + b + 9) - (9a + b + 1)| = 8|1 - a| \geq 4.$$

When $\dfrac{1}{2} < a < \dfrac{3}{2}$, we have $\dfrac{3}{\sqrt{a}} \in [1, 9]$. Further discussion can be divided into two cases:

If $\dfrac{1}{2} < a \leq 1$, then

$$\left| f(1) - f\left(\dfrac{3}{\sqrt{a}}\right) \right| = |(a + b + 9) - (6\sqrt{a} + b)| = (3 - \sqrt{a})^2 \geq 4.$$

If $1 < a < \dfrac{3}{2}$, then

$$\left| f(9) - f\left(\dfrac{3}{\sqrt{a}}\right) \right| = |(9a + b + 1) - (6\sqrt{a} + b)| = (3\sqrt{a} - 1)^2 \geq 4.$$

To sum up, it is clear that there exists $u, v \in [1, 9]$ satisfying $|f(u) - f(v)| \geq 4$. This completes the proof.

Solution 2 By contradiction. Assume that for any $x \in [1, 9]$, $|f(x)| < 2$. Then we have

$$|f(1)| < 2, \quad |f(3)| < 2, \quad |f(9)| < 2.$$

It is easy to find

$$f(1) = a + b + 9, \qquad \qquad \text{①}$$

$$f(3) = 3a + b + 3, \qquad \qquad \text{②}$$

$$f(9) = 9a + b + 1. \qquad \qquad \text{③}$$

From ① and ②, we get $2a - 6 = f(3) - f(1)$; From ② and ③, we find $6a - 2 = f(9) - f(3)$.

By eliminating a from the above two equations, it follows that

$$f(3) - 4f(2) + 3f(1) = (6a - 2) - 3 \cdot (2a - 6) = 16.$$

However, $f(3) - 4f(2) + 3f(1) < 2 + 4 \cdot 2 + 3 \cdot 2 = 16$, a contradiction. Therefore, we complete the proof.   □

2. (40 marks) As shown in Fig. 2.1, in isosceles triangle $\triangle ABC$, $AB = AC$, point D on side AC and point E on the extension of BC satisfy $\dfrac{AD}{DC} = \dfrac{BC}{2CE}$. Circle ω with diameter AB intersects line segment DE at point F.

Prove that B, C, F, D are concyclic.

Solution   As shown in Fig. 2.2, take the midpoint of BC, denoted by H. Then from $AB = AC$, we get $AH \perp BC$, and thus H is on circle ω.

Extend FD to G such that $AG // BC$. By the given condition, it follows that $\dfrac{AG}{CE} = \dfrac{AD}{DC} = \dfrac{BC}{2CE}$. Therefore,

$$AG = \frac{1}{2}BC = BH = HC.$$

Then $AGBH$ is a rectangle and $AGHC$ is a parallelogram.

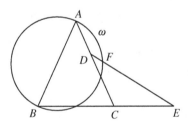

Fig. 2.1

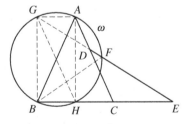

Fig. 2.2

As $AGBH$ is a rectangle, G is also on circle ω, and thus $\angle HGF = \angle HBF$.

As $AGHC$ is a parallelogram, then from $AC//GH$ we get $\angle CDF = \angle HGF$.

Therefore, $\angle CDF = \angle HBF = \angle CBF$.

Consequently, B, C, F, D are concyclic. $\qquad\square$

3 (50 marks) Given set $A = \{1, 2, \ldots, n\}$, X, Y are non-empty subsets of A ($X = Y$ is allowed). Denote the maximum element in X and the minimum element in Y by $\max X$, $\min Y$, respectively. Find the number of ordered set pairs (X, Y) that satisfy $\max X > \min Y$.

Solution First we calculate the number of ordered set pairs (X, Y) that satisfy $\max X \leq \min Y$. For a given $m = \max X$, set X is the union of any subset of set $\{1, 2, \ldots, m-1\}$ and $\{m\}$, so there are 2^{m-1} ways of selection. Since $\min Y \geq M$, thus Y is any non-empty subset of $\{m, m+1, \ldots, n\}$, and there are $2^{n+1-m} - 1$ ways of selection.

Therefore, the number of ordered set pairs (X, Y) that satisfy $\max X \leq \min Y$ is

$$\sum_{m=1}^{n} 2^{m-1}(2^{n+1-m} - 1) = \sum_{m=1}^{n} 2^n - \sum_{m=1}^{n} 2^{m-1} = n \cdot 2^n - 2^n + 1.$$

Since the number of ordered set pairs (X, Y) is $(2^n - 1) \cdot (2^n - 1) = (2^n - 1)^2$, then the number of ordered set pairs (X, Y) that satisfy $\max X > \min Y$ is

$$(2^n - 1)^2 - n \cdot 2^n + 2^n - 1 = 2^{2n} - 2^n(n+1). \qquad\square$$

4 (50 marks) Given integer $a \geq 2$, prove that for any positive integer n, there exist positive integers k such that n consecutive numbers $a^k + 1, a^k + 2, \ldots, a^k + n$ are all composite numbers.

Solution Let $i_1 < i_2 < \cdots < i_r$ be all the integers in $1, 2, \ldots, n$ that are mutually prime to a. Then for $1 \leq i \leq n$, $i \notin (i_1, i_2, \ldots, i_r)$, $a^k + i$ is not mutually prime to a and is greater than a, regardless of the value of positive integer k. Therefore, $a^k + i$ is a composite number.

For any $j = 1, 2, \ldots, r$, since $a + i_j > 1$, thus $a + i_j$ has prime factor p_j. Then we have $(p_j, a) = 1$. Therefore, by Fermat's little theorem we have

$$a^{p_j - 1} \equiv 1 \pmod{p_j}.$$

Now we take $k = (p_1 - 1)(p_2 - 1) \cdots (p_r - 1) + 1$.

For any $j = 1, 2, \ldots, r$, we have $k \equiv 1 \pmod{p_j - 1}$. Therefore,

$$a^k + i_j \equiv a + i_j \equiv 0 \pmod{p_j}.$$

Since $a^k + i_j > a + i_j \geq p_j$, thus $a^k + i_j$ is a composite number.

In summary, when $k = (p_1 - 1)(p_2 - 1) \cdots (p_r - 1) + 1$, $a^k + 1, a^k + 2, \ldots, a^k + n$ are all composite numbers. $\qquad\square$

China Mathematical Competition (Complementary Test)

2019

Test Paper A
(9:40 – 12:30; September 9, 2019)

1 (40 marks) As shown in Fig. 1.1, in acute $\triangle ABC$, M is the midpoint of side BC. Point P lies inside of $\triangle ABC$ such that AP bisects $\angle BAC$, and line MP intersects the circumscribed circles of $\triangle ABP, \triangle ACP$ at points D, E, respectively, which are different from P. Prove: if $DE = MP$, then $BC = 2BP$.

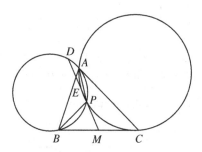

Fig. 1.1

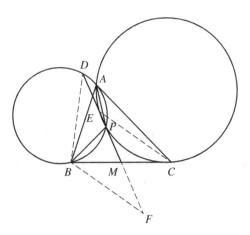

Fig. 1.2

Solution Extend PM to point F such that $MF = ME$. Join BF, BD, CE.

By the given condition, we obtain

$$\angle BDP = \angle BAP = \angle CAP = \angle CEP = \angle CEM.$$

Since $BM = CM$ and $EM = FM$, then $BF = CE$ and $BF // CE$.

Thus $\angle F = \angle CEM = \angle BDP$, and hence $BD = BF$.

For $DE = MP$, we have $DP = EM = FM$.

Therefore, in isosceles $\triangle BDF$, by symmetry we have $BP = BM$. Consequently, $BC = 2BM = 2BP$.

The proof is completed. □

2 (40 marks) Suppose integers $a_1, a_2, \ldots, a_{2019}$ satisfy $1 = a_1 \leq a_2 \leq \cdots \leq a_{2019} = 99$.

Denote $f = (a_1^2 + a_2^2 + \cdots + a_{2019}^2) - (a_1 a_3 + a_2 a_4 + a_3 a_5 + \cdots + a_{2017} a_{2019})$.

Find the minimum f_0 of f. And determine the number of arrays $(a_1, a_2, \ldots, a_{2019})$ such that $f = f_0$ is satisfied.

Solution From the given condition, we have

$$2f = a_1^2 + a_2^2 + a_{2018}^2 + a_{2019}^2 + \sum_{i=1}^{2017} (a_{i+2} - a_i)^2. \qquad \textcircled{1}$$

Since a_1, a_2 and $a_{i+2} - a_i$ $(i = 1, 2, \ldots, 2016)$ are all nonnegative integers, thus $a_1^2 \geq a_1, a_2^2 \geq a_2$ and $(a_{i+2} - a_i)^2 \geq a_{i+2} - a_i (i = 1, 2, \ldots, 2016)$.

Therefore,

$$a_1^2 + a_2^2 + \sum_{i=1}^{2016} (a_{i+2} - a_i)^2 \geq a_1 + a_2 + \sum_{i=1}^{2016} (a_{i+2} - a_i)$$

$$= a_{2017} + a_{2018}. \tag{2}$$

From ① and ②, we find

$$2f \geq a_{2017} + a_{2018} + (a_{2019} - a_{2017})^2 + a_{2018}^2 + a_{2019}^2,$$

Combining $a_{2019} = 99$ and $a_{2018} \geq a_{2017} > 0$, it follows that

$$f \geq \frac{1}{2} \left(2a_{2017} + (99 - a_{2017})^2 + a_{2017}^2 + 99^2 \right)$$

$$= (a_{2017} - 49)^2 + 7400$$

$$\geq 7400. \tag{3}$$

On the other hand, let

$$a_1 = a_2 = \cdots = a_{1920} = 1,$$

$$a_{1920+2k-1} = a_{1920+2k} = k(k = 1, 2, \ldots, 49),$$

$$a_{2019} = 99.$$

It can be verified that all the above inequalities can take the equal sign, and hence the minimum of f is $f_0 = 7400$.

The condition for the inequalities in ③ taking the equal sign is: $a_{2017} = a_{2018} = 49$, and the inequality in ② takes the equal sign, namely $a_1 = a_2 = 1$, $a_{i+2} - a_i \in \{0, 1\}$ ($i = 1, 2, \ldots, 2016$).

Therefore, $1 = a_1 \leq a_2 \leq \cdots \leq a_{2018} = 49$, and there are at least two elements of $a_1, a_2, \ldots, a_{2018}$ is equal to k for each $k(1 \leq k \leq 49)$. It is easy to verify that this is also the sufficient condition for ③ taking the equal sign.

For each k $(1 \leq k \leq 49)$, let the number of elements in $a_1, a_2, \ldots, a_{2018}$ that are equal to k be denoted by $1 + n_k$, then n_k is a positive integer and

$$(1 + n_1) + (1 + n_2) + \cdots + (1 + n_{49}) = 2018,$$

and that is

$$n_1 + n_2 + \cdots + n_{49} = 1969.$$

The number of positive integer solutions $(n_1, n_2, \ldots, n_{49})$ of the above equation is C_{1968}^{48}, and each solution uniquely corresponds to the array

$(a_1, a_2, \ldots, a_{2019})$ such that ④ takes the equal sign. Therefore the number of arrays $(a_1, a_2, \ldots, a_{2019})$ such that $f = f_0$ is C_{1968}^{48}.   □

③ (50 marks) Given integer m with $|m| \geq 2$, integer sequence $a_1, a_2, \ldots$ satisfies: a_1, a_2 are not both zero, and $a_{n+2} = a_{n+1} - ma_n$ holds for any positive integer n.

    Prove: if there exists integers $r, s(r > s \geq 2)$ such that $a_r = a_s = a_1$, then $r - s \geq |m|$.

Solution   Without loss of generality, assume a_1, a_2 are coprime. (Otherwise, if $(a_1, a_2) = d > 1$, then $\dfrac{a_1}{d}$ and $\dfrac{a_2}{d}$ are coprime, and we can substitute $a_1, a_2, a_3, \ldots$ for $\dfrac{a_1}{d}, \dfrac{a_2}{d}, \dfrac{a_3}{d}, \ldots$, while the conditions and conclusion remain unchanged.)

It is known from the recurrence relation of sequence that

$$a_2 \equiv a_3 \equiv a_4 \equiv \cdots \pmod{|m|}. \qquad ①$$

We will prove, for any integer $n \geq 3$,

$$a_n \equiv a_2 - (a_1 + (n-3)a_2)m \pmod{m^2}. \qquad ②$$

Indeed, it is immediately obvious that ② holds for $n = 3$. Assume ② holds for $n = k$ (where k is an integer greater than 2). From ①, we have $ma_{k-1} \equiv ma_2 \pmod{m^2}$. By the assumption and the principle of mathematical induction, it follows that

$$a_{k+1} = a_k - ma_{k-1} \equiv a_2 - (a_1 + (k-3)a_2)m - ma_2$$
$$\equiv a_2 - (a_1 + (k-2)a_2) \pmod{m^2}.$$

So ② also holds for $n = k + 1$, and then holds for any integer $n \geq 3$. It should be noted that, when $a_1 = a_2$, ② also holds for $n = 2$.

Suppose integers $r, s(r > s \geq 2)$ satisfy $a_r = a_s = a_1$. If $a_1 = a_2$, as ② holds for $n \geq 2$, then

$$a_2 - (a_1 + (r-3)a_2)m \equiv a_r = a_s \equiv a_2 - (a_1 + (s-3)a_2)m \pmod{m^2},$$

namely $a_1 + (r-3)a_2 \equiv a_1 + (s-3)a_2 \pmod{|m|}$, implying

$$(r-s)a_2 \equiv 0 \pmod{|m|}. \qquad ③$$

If $a_1 \neq a_2$, then $a_r = a_s = a_1 \neq a_2$, thus $r > s \geq 3$. Since ② holds for $n \geq 3$, similarly we find that ③ is also satisfied.

Now we prove that a_2, m are coprime. By contradiction, assume there exists a common prime factor p between a_2 and m. From ① we learn that p is the common factor of $a_2, a_3, a_4, \ldots$. But a_1, a_2 are coprime. Thus $p \nmid a_1$, which is in contradiction to $a_r = a_s = a_1$.

Therefore, $r - s \equiv 0 \pmod{|m|}$ from ③.

Since $r > s$, it follows that $r - s \geq |m|$. □

④ (50 marks) Suppose V is a set of 2019 points in space, in which any four points are not coplanar. Some points in V are joined by segments, which form a set of segments denoted by E. Find the least positive integer n satisfying: if E contains at least n elements, then E must have 908 2-element subsets, such that the two segments in each 2-element subset have common end points, and the intersection set of any two 2-element subsets is null.

Solution For the sake of convenience, we call two adjacent edges in a graph constitute an *angle*. Firstly, we shall prove a lemma.

Lemma *Suppose $G = (V, E)$ is a simple graph and G is connected, then G has $\left[\dfrac{|E|}{2} \right]$ angles with no common edge between each angle pair, where $[\alpha]$ denotes the integer part of real number α.*

Proof of the lemma We use mathematical induction. When $|E| = 0, 1, 2, 3$, the conclusion is obviously correct. In the following we assume $|E| \geq 4$ and the conclusion is established for smaller $|E|$. We only need to prove: choose two edges a, b in G that form an angle, after deleting a, b in G, there exists a connected component in the rest graph with $|E| - 2$ edges. Applying the induction method to this connected component will lead to the conclusion.

Consider the longest path $P : v_1 v_2 \ldots v_k$ in G, where $v_1, v_2, \ldots, v_k$ are different vertices. Since G is connected, thus $k \geq 3$.

Case 1: $\deg(v_1) \geq 2$. Since P is the longest path, the adjacent vertices of v_1 lies in $v_2, \ldots, v_k$. Suppose $v_1 v_i \in E$, where $3 \leq i \leq k$, then $\{v_1 v_2, v_1 v_i\}$ is an angle. Delete these two edges in E. If there is a third edge at v_1, then the rest graph is connected; if there are no other edges at v_1 besides the two ones deleted, then v_1 will become an isolated vertex while the rest vertices are still connected. In short, there exists a connected component in the rest graph with $|E| - 2$ edges.

Case 2: $\deg(v_1) = 1$, $\deg(v_2) = 2$. Then $\{v_1v_2, v_2v_3\}$ is an angle. Deleting these two edges, then v_1, v_2 become isolated vertices while the other vertices are connected to each other. Then there is a connected component with $|E| - 2$ edges.

Case 3: $\deg(v_1) = 1$, $\deg(v_2) \geq 3$, and v_2 is adjacent to some vertex in $v_4, \ldots, v_k$. Then $\{v_1v_2, v_2v_3\}$ is an angle. Deleting these two edges, v_1 will become an isolated vertex while the other vertices are connected to each other. Then the rest G has a connected component with $|E| - 2$ edges.

Case 4: $\deg(v_1) = 1$, $\deg(v_2) \geq 3$, and v_2 is adjacent to some vertex $u \notin \{v_1, v_3, \ldots, v_k\}$. Since P is the longest path, the adjacent vertices of u all lie in $v_2, \ldots, v_k$. Then $\{v_1v_2, v_2u\}$ is an angle, and by deleting these two edges, v_1 becomes an isolated vertex. If there is just one edge uv_2 at u, then u is also an isolated vertex after deleting the edges mentioned while the other points are connected to each other. If there are other edges uv_i $(3 \leq i \leq k)$ at u, then the other vertices besides v_1 are connected to each other after deleting the edges mentioned. In a word, there exists a connected component in the rest graph with $|E| - 2$ edges.

The proof of the lemma is completed.

Let us go back to the original problem. V and E can be viewed as a graph $G = (V, E)$.

Firstly, we will prove that $n \geq 2795$.

Let $V = \{v_1, v_2, \ldots, v_{2019}\}$. In $v_1, v_2, \ldots, v_{61}$, firstly join the vertices in pairs, and then delete 15 of the edges (e.g. $v_1v_2, v_1v_3, \ldots, v_1v_{16}$). The number of total edges is $C_{61}^2 - 15 = 1815$. The graph comprised by these 61 vertices is then connected. Next divide the rest $2019 - 61 = 1958$ vertices into 979 pairs with the two vertices in each pair are connected by a segment, so there are $1815 + 979 = 2794$ segments in graph G. From the above discussion we can see that every angle in G must be comprised by the sides connected by $v_1, v_2, \ldots, v_{61}$, therefore there are at most $\left\lceil \dfrac{1815}{2} \right\rceil = 907$ angles with no common sides. Therefore, in order to satisfy the given condition, n is not less than 2795.

On the other hand, if $|E| \geq 2795$, we can delete several edges and only consider the case $|E| = 2795$.

Suppose G has k connected components with $m_1, \ldots, m_k$ vertices and $e_1, \ldots, e_k$ edges, respectively. In the following we shall prove that there are at most 979 odd numbers among $e_1, \ldots, e_k$.

We shall make use of *proof by contradiction*. Assume there are at least 980 odd numbers among $e_1, \ldots, e_k$. Since $e_1 + \cdots + e_k = 2795$ is odd, thus $e_1, \ldots, e_k$ has at least odd numbers, and therefore $k \geq 981$. Without loss of generality, we assume $e_1, e_2, \ldots, e_{981}$ are all odd. It is obvious that $m_1, m_2, \ldots, m_{981} \geq 2$.

Let $m = m_{981} + \cdots + m_k \geq 2$. Then $C_{m_i}^2 \geq e_i$ $(1 \leq i \leq 980)$ and $C_m^2 \geq e_{981} + \cdots + e_k$. Hence

$$2795 = \sum_{i=1}^{k} e_i \leq C_m^2 + \sum_{i=1}^{980} C_{m_i}^2. \qquad ①$$

By using of the convex of combinatorial numbers, we have $C_x^2 + C_y^2 \leq C_{x+1}^2 + C_{y-1}^2$, for $x \geq y \geq 3$. $C_m^2 + \sum_{i=1}^{980} C_{m_i}^2$ has the maximum value when $m_1, \ldots, m_{980}, m$ consists of 980 integers 2 and one 59. Hence

$$C_m^2 + \sum_{i=1}^{980} C_{m_i}^2 \leq C_{59}^2 + 980 C_2^2 = 2691 < 2795,$$

inconsistent with ①. Hence there are at most 979 odd numbers among $e_1, \ldots, e_k$.

Applying the lemma to each connected component, we observe that G has N angles with no common sides, where

$$N = \sum_{i=1}^{k} \left[\frac{e_i}{2} \right] \geq \frac{1}{2} \left(\sum_{i=1}^{k} e_i - 979 \right) = \frac{1}{2}(2795 - 979) = 908.$$

To sum up, the minimum value of n is 2795. □

Test Paper B
(9:40 – 12:30; September 8, 2019)

1 (40 marks) Suppose positive real numbers $a_1, a_2, \ldots, a_{100}$ satisfy $a_i \geq a_{101-i}$ $(i = 1, 2, \ldots, 50)$. Denote $x_k = \dfrac{k a_{k+1}}{a_1 + a_2 + \cdots + a_k}$ $(k = 1, 2, \ldots, 99)$.

Prove that $x_1 x_2^2 \ldots x_{99}^{99} \leq 1$.

Solution   Note that $a_1, a_2, \ldots, a_{100} > 0$. For $k = 1, 2, \ldots, 99$, by using of the inequality of arithmetic and geometric means, we find

$$0 < \left(\frac{k}{a_1 + a_2 + \cdots + a_k} \right)^k \leq \frac{1}{a_1 a_2 \ldots a_k},$$

and hence

$$x_1 x_2^2 \ldots x_{99}^{99} = \prod_{k=1}^{99} a_{k+1}^k \left(\frac{k}{a_1 + a_2 + \cdots + a_k} \right)^k \leq \prod_{k=1}^{99} \frac{a_{k+1}^k}{a_1 a_2 \cdots a_k} \qquad \text{(1)}$$

Denote the right side of ① by T. Then for any $i = 1, 2, \ldots, 100$, the power of a_i in the numerator of T is $i - 1$ and that in its denominator is $100 - i$. Hence

$$T = \prod_{i=1}^{100} a_i^{2i-101} = \prod_{i=1}^{50} a_i^{2i-101} a_{100-i}^{2(101-i)-101} = \prod_{i=1}^{50} \left(\frac{a_{100-i}}{a_i} \right)^{101-2i}.$$

As $0 < a_{101-i} \leq a_i$ ($i = 1, 2, \ldots, 50$), $T \leq 1$. Combing with ① yields

$$x_1 x_2^2 \ldots x_{99}^{99} \leq T \leq 1. \qquad \square$$

② (40 marks) Find all the positive integers n satisfying the following conditions:

(1) n has at least 4 positive factors;

(2) If $d_1 < d_2 < \cdots < d_k$ are all the positive factors of n, then $d_2 - d_1, d_3 - d_2, \ldots, d_k - d_{k-1}$ is a geometric sequence.

Solution   By the given condition, we know that $k \geq 4$ and $\dfrac{d_3 - d_2}{d_2 - d_1} = \dfrac{d_k - d_{k-1}}{d_{k-1} - d_{k-2}}$.

It is easy to find that $d_1 = 1$, $d_k = n$, $d_{k-1} = \dfrac{n}{d_2}$, $d_{k-2} = \dfrac{n}{d_3}$. Putting into the above formula, we obtain

$$\frac{d_3 - d_2}{d_2 - 1} = \frac{n - \dfrac{n}{d_2}}{\dfrac{n}{d_2} - \dfrac{n}{d_3}}.$$

Simplifying it yields

$$(d_3 - d_2)^2 = (d_2 - 1)^2 d_3.$$

From this we find d_3 is a perfect square. Since $d_2 = p$ is the least prime factor of n, then the only case is $d_3 = p^2$.

Therefore, $d_2 - d_1, d_3 - d_2, \ldots, d_k - d_{k-1}$ is $p-1, p^2-p, p^3-p^2, \ldots, p^{k-1} - p^{k-2}$, namely $d_1, d_2, d_3, \ldots, d_k$ is $1, p, p^2, \ldots, p^{k-1}$, and the corresponding n is p^{k-1}.

To sum up, the desired n satisfying the given condition are positive integers having the form p^a, where p is a prime number and integer $a \geq 3$. $\qquad\square$

3 (50 marks) As shown in Fig. 3.1, points A, B, C, D, E are arranged in order in a straight line with $BC = CD = \sqrt{AB \cdot DE}$. Point P lies outside the line such that $PB = PD$. Points K, L lies on segments PB, PD, respectively, such that KC bisects $\angle BKE$ and LC bisects $\angle ALD$. Prove that A, K, L, E are concyclic.

Solution Let $AB = 1$, $BC = CD = t(t > 0)$. By the given condition we have $DE = t^2$.

As shown in Fig. 3.2, note that $\angle BKE < \angle ABK = \angle PDE < 180° - \angle DEK$. Take a point A' on the extension of CB such that

$$\angle A'KE = \angle ABK = \angle A'BK.$$

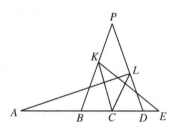

Fig. 3.1

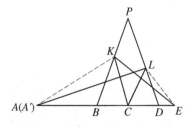

Fig. 3.2

Then we have $\triangle A'BK \sim \triangle A'KE$. Hence

$$\frac{A'B}{A'K} = \frac{A'K}{A'E} = \frac{BK}{KE}.$$

Since KC bisects $\angle BKE$, it follows that

$$\frac{BK}{KE} = \frac{BC}{CE} = \frac{t}{t+t^2} = \frac{1}{1+t}.$$

And hence

$$\frac{A'B}{A'E} = \frac{A'B}{A'K} \cdot \frac{A'K}{A'E} = \left(\frac{BK}{KE}\right)^2$$

$$= \frac{1}{1+2t+t^2} = \frac{AB}{AE}.$$

Subtract both sides of the above equation by 1, we get $\dfrac{BE}{A'E} = \dfrac{BE}{AE}$, and then $A' = A$. Therefore,

$$\angle AKE = \angle A'KE = \angle ABK.$$

Similarly, we obtain $\angle ALE = \angle EDL$.

Since $\angle ABK = \angle EDL$, then $\angle AKE = \angle ALE$. Accordingly, A, K, L, E are concyclic. $\qquad\square$

4 (50 marks) Given a convex 2019-sided polygon, assign one of three colors — red, yellow, and blue — to each edge, such that the number of the edges with each color is equally 673. Prove that: in this polygon, we can make 2016 diagonals that do not intersect with each other, to divide the polygon into 2017 triangles, and then assign one of the three colors to each diagonal such that for each triangle the colors of its three sides are either the same or different from each other.

Solution For $n \geq 5$, we shall make a strengthened proposition: if we assign one of the three colors a, b, c to each edge of a convex n-sided polygon, such that each color is assigned to at least one edge, then we can make the triangles in this polygon, to satisfy the requirements.

When $n = 5$, if the numbers of the edges of the three colors are 1, 1, 3, respectively, by symmetry we only need to consider two cases and can make the division of triangles shown in Fig. 4.1.

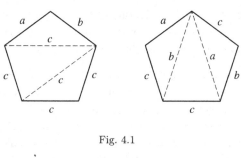

Fig. 4.1

Fig. 4.2

If the number of the edges of the three colors are 1, 2, 2, respectively, by symmetry we only need to consider three cases and can make the division of triangles shown in Fig. 4.1.

Suppose the conclusion holds for $n(n \geq 5)$. Then we consider the situation for $n + 1$. Denote the convex $n + 1$-sided polygon by $A_1 A_2 \ldots A_{n+1}$.

Case 1: Two colors are each assigned to only one edge, respectively. Without loss of generality, assume colors a, b are each assigned to one edge, respectively. Since $n + 1 \geq 6$, then there exist two successive edges assigned with color c, which we denote by $A_n A_{n+1}$, $A_{n+1} A_1$. Make the diagonal $A_1 A_n$ and assign it with color c, then the three sides of triangle $A_n A_{n+1} A_1$ are of the same color. Then for convex n-sided polygon $A_1 A_2 \ldots A_n$, each color is assigned to at least one edge. By induction, it follows that we can make the division of triangles satisfying the requirements.

Case 2: One of the three colors is assigned to only one edge, while the other two colors are each assigned to at least two edges, respectively. Without loss of generality, suppose colors a is assigned to only one edge. Then we can choose two adjacent edges that are not assigned with color a, denoting them by $A_n A_{n+1}$, $A_{n+1} A_1$. Make diagonal $A_1 A_n$, and then $A_1 A_n$ has the unique coloring, such that the colors of the three sides of triangle $A_n A_{n+1} A_1$ are either of the same or different from each other.

Therefore, for convex n-sided polygon $A_1 A_2 \ldots A_n$, each color is assigned to at least one edge, respectively. By induction, it follows that we can make the division of triangles satisfying the requirements.

Case 3: Each color is assigned to at least two edges, respectively. Make diagonal $A_1 A_n$, and then $A_1 A_n$ has the unique coloring, such that the colors of the three sides of triangle $A_n A_{n+1} A_1$ are either of the same or different from each other. Therefore, for convex n-sided polygon $A_1 A_2 \ldots A_n$, each color is assigned to at least one edge. By induction, it follows that we can make the division of triangles satisfying the requirements.

Combining the above results, we see that the conclusion is established in the situation for $n + 1$.

By the principle of mathematical induction, it follows that the proof is completed. □

China Mathematical Olympiad

2018 (Chengdu, Sichuan)

2018 China Mathematical Olympiad (also named the 34th National Mathematics Winter Camp for Middle School Students), was organized by the China Mathematical Olympiad Committee and hosted by the Sichuan Mathematical Society and Chengdu Seventh Middle School. It was held in Chengdu from November 12 to 16, 2018, with 35 teams from 31 provinces, municipalities and autonomous regions in China's mainland, Hong Kong Special Administrative Region of China, Macao Special Administrative Region of China, Singapore and Russia, with a total of 395 secondary school students. As a result of the competition, 124 students won the first prize, 156 students won the second prize and 87 students won the third prize.

60 students were selected from this competition to form the national training team.

The members of the main examination committee of this winter camp are:

Zhou Qing (East China Normal University);

Xiong Bin (East China Normal University);

Yao Yijun (Fudan University);

Qu Zhenhua (East China Normal University);

He Yijie (East China Normal University);

Wang Bin (Chinese Academy of Sciences Academy of Mathematics and Systems Science);

Wang Xinmao (University of Science and Technology of China);

Li Weigu (Peking University);

Jin Long (Tsinghua University);
Lai Li (Tsinghua University);
Li Ting (Sichuan University);
Fu Yunhao (Southern University of Science and Technology);
Ji Chungang (Nanjing Normal University);
Zhang Sihui (University of Shanghai for Science and Technology);
Wang guozhen (Fudan University);
Leng Fusheng (Chinese Academy of Sciences Academy of Mathematics and Systems Science);
Lin Tianqi (East China Normal University).

First Day
(8:00 – 12:30; November 14, 2018)

1 Suppose real numbers $a, b, c, d, e \geq -1$ and $a+b+c+d+e = 5$. Find the minimum and maximum values of $S = (a + b)(b + c)(c + d)(d + e)(e + a)$. (Contributed by Xiong Bin)

Solution First, we find the minimum value of S. We only need to consider the case $S < 0$.

(1) If four of $a + b$, $b + c$, $c + d$, $d + e$, $e + a$ are positive and one is negative, then by symmetry, we suppose $a + b < 0$. Noticing that $a + b + c + d + e = 5$ and $a + b \geq -2$, by the inequality of arithmetic and geometric means, we have

$$(b + c)(c + d)(d + e)(e + a) \leq \left[\frac{(b + c) + (c + d) + (d + e) + (e + a)}{4} \right]^4$$
$$= \left(\frac{10 - a - b}{4} \right)^4 \leq 3^4 = 81,$$

and hence $S \geq -2 \times 81 = -162$.

(2) If two of $a + b$, $b + c$, $c + d$, $d + e$, $e + a$ are positive and three are negative, since

$$-2 \leq a + b = 5 - c - d - e \leq 8,$$

in the same way we get $-2 \leq b + c, c + d, d + e, e + a \leq 8$. Therefore, $S \geq (-2)^3 \times 8^2 = -512$.

(3) If $a+b, b+c, c+d, d+e, e+a$ are all negative, i.e. $a+b+c+d+e < 0$. This contradicts the given condition, so this is impossible.

In summary, $S \geq -512$. When $a = b = c = d = -1$ and $e = 9$, S takes the minimum value of -512.

Next, we find the maximum value of S. We only need to consider the case $S > 0$.

(1) If $a + b, b + c, c + d, d + e, e + a$ are all positive, by the inequality of arithmetic and geometric means, we have

$$S \leq \left[\frac{(a+b) + (b+c) + (c+d) + (d+e) + (e+a)}{5} \right]^5 = 32.$$

(2) If three of $a+b, b+c, c+d, d+e, e+a$ are positive and two are negative, then by symmetry, we only need to consider the following two cases.

  (i) $a + b, b + c < 0$ and $c + d, d + e, e + a > 0$. Then we have

$$-2 \leq a + b < 0, -2 \leq b + c < 0, 0 < d + e \leq 8,$$

$$0 < (c+d)(e+a) \leq \left(\frac{c+d+e+a}{2} \right)^2 = \left(\frac{5-b}{2} \right)^2 \leq 9,$$

  and hence $S \leq (-2) \times (-2) \times 8 \times 9 = 288$.

  (ii) $a + b, c + d < 0$ and $b + c, d + e, e + a > 0$. Notice that

$$e = 5 - a - b - c - d \leq 7 - b - c < 7.$$

By the inequality of arithmetic and geometric means, it follows that

$$S \leq \left[\frac{(-a - b) + (b + c) + (-c - d) + (d + e) + (e + a)}{5} \right]^5$$

$$= \left(\frac{2e}{5} \right)^2 < 3^5 < 288.$$

(3) If one of $a + b, b + c, c + d, d + e, e + a$ is positive and four are negative, then by symmetry, we suppose $a + b > 0$. Therefore,

$$0 > (b + c) + (c + d) + (e + a) = 5 + c \geq 4.$$

However, this is in contradiction to the given condition and cannot be valid.

In summary, $S \leq 288$. When $a = b = c = -1$ and $d = e = 4$, S takes the maximum value of 288. $\qquad \square$

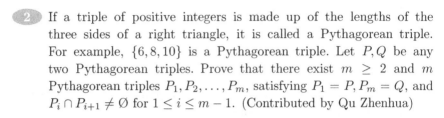 If a triple of positive integers is made up of the lengths of the three sides of a right triangle, it is called a Pythagorean triple. For example, $\{6, 8, 10\}$ is a Pythagorean triple. Let P, Q be any two Pythagorean triples. Prove that there exist $m \geq 2$ and m Pythagorean triples $P_1, P_2, \ldots, P_m$, satisfying $P_1 = P, P_m = Q$, and $P_i \cap P_{i+1} \neq \varnothing$ for $1 \leq i \leq m - 1$. (Contributed by Qu Zhenhua)

Solution　We first prove the following lemma:

Lemma　*For every integer $n \geq 3$, there exists a Pythagorean triple containing n.*

Proof of lemma　We will complete the proof of lemma by applying induction on n. Since $\{3, 4, 5\}$ is a Pythagorean triple, the conclusion holds for $n = 3, 4, 5$. We assume that $n \geq 6$ and that the conclusion holds for all $3 \leq k < n$. If n is even, let $n = 2k$, then $3 \leq k < n$. By induction hypothesis, there exists a Pythagorean triple $\{k, a, b\}$, so $\{n, 2a, 2b\}$ is also a Pythagorean triple. If n is odd, it is easy to see that $\left\{ n, \dfrac{1}{2}(n^2 - 1), \dfrac{1}{2}(n^2 + 1) \right\}$ is a Pythagorean triple. The lemma is the proved.

If two Pythagorean triples P, Q satisfy the conclusion, we denote it by $P \sim Q$. It is an equivalence relation. Then we only need to prove that for any Pythagorean triple P, we have $P \sim \{3, 4, 5\}$.

For a Pythagorean triple $P = \{a, b, c\}$ and positive integer k, let $kP = \{ka, kb, kc\}$. Then kP is also is a Pythagorean triple. It is easy to know that if $P \sim Q$, then $kP \sim kQ$.

We will prove by induction on the smallest element of P that $P \sim \{3, 4, 5\}$.

If $3 \leq \min P \leq 5$, it is clear that $P \sim \{3, 4, 5\}$. Note that

$$\{6, 8, 10\} \sim \{8, 15, 17\} \sim \{9, 12, 15\} \sim \{5, 12, 13\} \sim \{3, 4, 5\}.$$

Assume that $\min P = n \geq 6$ and the conclusion holds for $3 \leq \min P < n$.

If n is even, let $n = 2k$, and then $3 \leq k < n$. By the lemma, there exists a Pythagorean triple set P' containing k. Since $\min P' \leq k < n$, by the induction hypothesis, we know that $P' \sim \{3, 4, 5\}$, and thus

$$P \sim 2P' \sim 2\{3, 4, 5\} = \{6, 8, 10\} \sim \{3, 4, 5\}.$$

If n is odd, then $P \sim \left\{ n, \dfrac{1}{2}(n - 1)(n + 1), \dfrac{1}{2}\left(n^2 + 1\right) \right\}$. By the lemma, there exist a Pythagoras triple Q containing $\dfrac{1}{2}(n + 1)$ and a Pythagoras triple R containing $n - 1$. Since $3 \le \frac{1}{2}(n + 1) < n$ and $3 \le n - 1 < n$, by induction hypothesis we have $Q \sim \{3, 4, 5\}$, $R \sim \{3, 4, 5\}$, and thus

$$\left\{ n, \frac{1}{2}(n - 1)(n + 1), \frac{1}{2}(n^2 + 1) \right\} \sim (n - 1)Q \sim (n - 1)\{3, 4, 5\}$$

$$\sim 3R \sim 3\{3, 4, 5\} = \{9, 12, 15\} \sim \{3, 4, 5\}.$$

Therefore, by induction we know that for any Pythagorean triple $P, P \sim \{3, 4, 5\}$. Consequently, the conclusion is proved. $\square$

3 As shown in Fig. 3.1, $\triangle ABC$ is inscribed in circle $\odot O$ with $AB < AC$. Point D is on the bisector of $\angle BAC$ and point E is on side BC such that $OE // AD, DE \perp BC$. Point K is on the extension of EB such that $EA = EK$. The circle passing through points A, K, D intersects the extension of BC at point P and intersects $\odot O$ at A and another point Q.

Prove that line PQ is tangent to $\odot O$. (Contributed by He Yijie)

Solution 1 As shown in Fig. 3.2, suppose AD intersects arc BC of circle $\odot O$ at M, and thus M is the midpoint of BC. Connecting OA, OM,

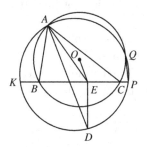

Fig. 3.1

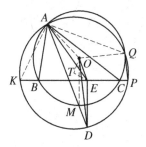

Fig. 3.2

we have $OM \perp BC$. Since $DE \perp BC$, then $OM // DE$. Furthermore, since $OE // MD$, thus quadrilateral $OMDE$ is a parallelogram, and hence $DE = OM = OA$. Therefore, quadrilateral $AOED$ is a isosceles trapezoid.

Connect AK, AQ, OQ, and let T be the center of $\odot AKDPQ$. Then T lies on the perpendicular bisectors of AK, AQ, AD, simultaneously. Connect TO, TE. Since the axis of symmetry of isosceles trapezoid $AOED$ passes through point T, thus by symmetry, we have $\angle TOA = \angle TED$.

From $EK = EA$, we know that ET is the perpendicular bisector of AK. From $OA = OQ$, we learn that OT is the perpendicular bisector of AQ. Therefore,

$$\angle AQO = \angle QAO = \angle TOA - 90° = \angle TED - 90°$$
$$= \angle TEK = 90° - \angle AKE.$$

And then

$$\angle OQP = \angle AQP - \angle AQO = (180° - \angle AKP) - (90° - \angle AKE) = 90°.$$

Therefore, PQ is tangent to $\odot O$. $\qquad \square$

Solution 2 Connect OA, OD. Take point Q' on $\odot O$ such that $\angle Q'OD = 90°$, and Q, D are on opposite sides of line AO. In the following we will prove that $Q' = Q$.

Connect $AK, KD, AQ', Q'D$. Suppose AD intersects arc of circle $\odot O$ at M, and thus M is the midpoint of arc $\overset{\frown}{BC}$ and $OM \perp BC$. Since $DE \perp BC$, then $OM // DE$. Furthermore, since $OE // MD$, thus the quadrilateral $OMDE$ is parallelogram, and hence $DE = OM = OA$. Therefore

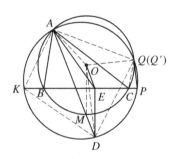

Fig. 3.3

quadrilateral $AOED$ is isosceles trapezoid, and then

$$DE = AO = OQ', EK = EA = OD.$$

Combining $\angle DEK = \angle Q'OD = 90°$, we learn that $\triangle DEK$ is congruent to $\triangle Q'OD$. Therefore,

$$\angle OQ'D + \angle DKE = \angle EDK + \angle DKE = 90°.$$

Furthermore, since

$$\angle AQ'O = 90° - \frac{1}{2}\angle AOQ' = 90° - \frac{1}{2}(270° - \angle AOD)$$

$$= \frac{1}{2}(\angle AOD - 90°) = \frac{1}{2}(\angle AED - 90°)$$

$$= \frac{1}{2}\angle AEK = 90° - \angle AKE,$$

thus

$$\angle AQ'D + \angle AKD = (\angle AQ'O + \angle AKE) + (\angle OQ'D + \angle DKE)$$

$$= 90° + 90° = 180°.$$

Consequently, A, K, D, Q' are concyclic. It is obvious that $Q' \neq A$, and thus $Q' = Q$. Since $\triangle DEK$ is congruent to $\triangle Q'OD$, thus there is $DK = DQ$, and hence $\angle KAD = \angle QAD$. And since $\angle BAD = \angle CAD$, thus $\angle KAB = \angle QAC$.

Connect QC and note that

$$\angle PQC = \angle BCQ - \angle KPQ = (180° - \angle BAQ) - (180° - \angle KAQ)$$

$$= \angle KAB = \angle QAC.$$

Therefore, PQ is tangent to $\odot O$. $\qquad\square$

Second Day
(8:00 – 12:30; November 15, 2018)

4 The following geometric figure is an ellipse with different lengths of major and minor axes.

(1) Prove that the rhombus circumscribing this ellipse with the smallest area is unique.

(2) Write down the process of making this rhombus using the method of ruler-and-compass construction. (Contributed by Yao Yijun)

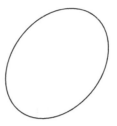

Fig. 4.1

Solution (1) In a coordinate system, we can set the center O of the ellipse as the origin and the axes of symmetry of the ellipse as the axes of coordinates. Then the equation of the ellipse is

$$\frac{x^2}{a^2} + \frac{y^2}{b^2} = 1, a > b > 0.$$

(i) According to the central symmetry property of an ellipse, the tangent points of the parallelogram circumscribing the ellipse on the opposite sides must be the endpoints of a diameter of the ellipse. Therefore, such a parallelogram is completely determined by two points $A(x_1, y_1)$, $B(x_1, y_1)$ on the ellipse that are not the endpoints of a diameter.

(ii) Note that the intersection of the tangents of the ellipse at points A and B is

$$P\left(\frac{y_2 - y_1}{x_1 y_2 - x_2 y_1} a^2, \frac{x_1 - x_2}{x_1 y_2 - x_2 y_1} b^2\right),$$

and the intersection of the tangents of the ellipse at points A and $-B$ is $Q\left(\dfrac{y_2 + y_1}{x_1 y_2 - x_2 y_1} a^2, \dfrac{-x_1 - x_2}{x_1 y_2 - x_2 y_1} b^2\right)$. This circumscribed parallelogram is a rhombus if and only if $\angle POQ$ is a right angle, and this is equivalent to

$$(y_2 - y_1)(y_1 + y_2) - (x_1 + x_2)(x_1 - x_2) = 0,$$

namely,

$$x_1^2 + y_1^2 = x_2^2 + y_2^2.$$

Combing

$$\frac{x_1^2}{a^2} + \frac{y_1^2}{b^2} = \frac{x_2^2}{a^2} + \frac{y_2^2}{b^2},$$

it follows that $x_1^2 = x_2^2, y_1^2 = y_2^2$, and thus A and B are symmetric about one of the coordinate axes. Therefore, the axis of symmetry of the rhombus circumscribing this ellipse must be either the major axis or the minor axis.

(iii) According to the above discussion about the circumscribed rhombus of the ellipse, make a stretching/shrinking transformation with coefficient $\dfrac{a}{b}$ in the y axis direction, and then the rhombus will become a circumscribed rhombus of a circle with radius a, its diagonals being coincide with the coordinate axes. The rhombus with the smallest area is a square and is unique. Therefore, the vertices of the rhombus with the smallest area are on the axes of symmetry of the ellipse, and the distances from the vertices to the center of the ellipse are $\sqrt{2}a$ and $\sqrt{2}b$, respectively.

(2) *Step* 1: make the center of symmetry and the axes of symmetry of the ellipse.

(i) To make the center of symmetry of the ellipse: Making two parallel chords of the ellipse, take their midpoints and make a line through them, so that the line segment intercepted by the ellipse is a diameter of the ellipse (It is obvious that the ellipse will become a circle after a stretching/shrinking transformation). The midpoint of this diameter is the center of symmetry of the ellipse.

(ii) To make the axes of symmetry of the ellipse: Taking the center of symmetry as the center, make a circle through a point on the ellipse, which intersects the latter at four points, so that get the two diameters of the ellipse, being symmetric about its two axes of symmetry, respectively. Then make the bisectors of the angles between the two diameters, to get the major and minor axes of the ellipse, respectively.

Step 2: After obtaining, in the first step, the axes of symmetry of the ellipse and the lengths of its semi-major and semi-minor axes a, b, respectively, take a point on the minor axis, whose distance to the center of the ellipse is a, and then its distance to one of the endpoints of the ellipse on the major axis is $\sqrt{2}a$. Then make two points on the major axis with the distances to the center of the ellipse are both $\sqrt{2}a$. Similarly, make two points on the minor axis, so that their distances of to the center of the ellipse are both $\sqrt{2}b$. Then the quadrilateral with these four points as the vertices is the rhombus required. □

5 Given positive integer n, fill every cell in the $n \times n$ grid with an integer each. Do the following with this grid: select a cell and add 1 to each of the other $2n - 1$ cells in the row and column where this cell is located.

Find the maximum integer $N = N(n)$ such that no matter how the numbers are distributed in the grid at the beginning, there are at least N even numbers in the grid after a series of the above operations. (Contributed by Wang Xinmao)

Solution   Let us choose a cell located in row i and column j, and denote the operation in the question as $M(i, j)$. Note that the result of a series of operations is independent of their order.

When n is even, consider $2n - 1$ operations $M(p, q)$ ($p = i$ or $q = j$). These operations act $2n - 1$ times for the cell located in row i and column j, n times for each of the remaining cells in either row i or column j, and 2 times for each of the cells in neither row i nor column j. Thus, the result of these $2n - 1$ operations is to change the parity of the number of the cell in row i and column j, while keeping the parity of the numbers in the other cells unchanged. Therefore, there is always a series of operations that can be performed so that the numbers in the cells are all even, and the required $N = n^2$.

When n is odd, then by the result above, it is always possible to make the $(n-1)^2$ numbers in the cells of the upper left part of the gird to be even numbers after a series of operations. Now, if the number of odd numbers in the grid is not less than n, then operation $M(n, n)$ can make the number of odd numbers in the grid no more than $n - 1$. Therefore, the required $N \geq n^2 - n + 1$.

On the other hand, denote the sums of the numbers in each row of the gird by $s_1, \ldots, s_n$ and that of the numbers in each column by $t_1, \ldots, t_n$, respectively. An operation will change the parity of every $s_1, \cdots, s_n, t_1, \ldots, t_n$. At the beginning, let $s_1, \ldots, s_{n-1}$ be even and $t_1, \ldots, t_{n-1}$ be odd, for example, like the following gird

$$\begin{bmatrix} 0 & \cdots & 0 & 0 \\ \vdots & & \vdots & \vdots \\ 0 & \cdots & 0 & 0 \\ 1 & \cdots & 1 & 0 \end{bmatrix}.$$

After making odd times operations on this gird, $s_1, \ldots, s_{n-1}$ become odd numbers, so there is at least one odd number in each of the first $n - 1$

rows; after making even times operations, $t_1, \ldots, t_{n-1}$ are still odd numbers, so there is at least one odd number in each of the first $n - 1$ columns. Therefore, there are always at least $n - 1$ odd numbers, i.e. the number of the cells with even numbers $\leq n^2 - n + 1$ in the grid. Consequently, the required $N = n^2 - n + 1$. $\qquad\square$

6 Place 2018 points $P_1, P_2, \ldots, P_{2018}$ (they can be placed on the same position) on the interior or boundary of a regular pentagon. Find all placements such that

$$S = \sum_{1 \leq i < j \leq 2018} |P_i P_j|^2$$

takes the maximum value, and prove your conclusion. (Contributed by Wang Bin)

Solution The ratio of the side length to the diagonal of a regular pentagon is the golden mean $\lambda = \dfrac{\sqrt{5} - 1}{2}$. Let the radius of the circumscribed circle of the regular pentagon be 1 and the five vertices be exactly the fifth roots of unity on the complex plane. Since $\cos \dfrac{2\pi}{5} = \dfrac{\sqrt{5} - 1}{4} = \dfrac{\lambda}{2}, \cos \dfrac{4\pi}{5} = 2\cos^2 \dfrac{2\pi}{5} - 1 = \dfrac{-1 - \lambda}{2}$, it is easy to know that the square of the side length of the regular pentagon is $2 - \lambda$ and the square of the length of its diagonal is $3 + \lambda$.

In the Cartesian coordinate system, suppose the coordinates of point P_i is (x_i, y_i). When considering the maximum value of S, move only one of the points at a time, e.g. P_1, and fix the remaining 2017 points. Then consider S as a function of $P_1(x_1, y_1)$:

$$S = \sum_{1 \leq i < j \leq 2018} |P_i P_j|^2 = \sum_{1 \leq i < j \leq 2018} (x_i - x_j)^2 + (y_i - y_j)^2$$

$$= 2017 \left[(x_1 - a)^2 + (y_1 - b)^2 \right] + c.$$

where a, b, c are only related to the positions of $P_2, \ldots, P_{2018}$. At this point maximizing S means maximizing the distance between point P_1 and point (a, b). In a convex pentagon the point with the greatest distance to a given point must be taken at some vertex. Because if P_1 is not at a vertex, then we can move it to a vertex and make S to be strictly larger. In this way we can place all the 2018 points at the vertices after a finite number of moves.

Let there be m_i points placed on the point $\omega^i, i = 0, 1, 2, 3, 4$. Here

$$\omega = \cos \frac{2\pi}{5} + \sqrt{-1} \sin \frac{2\pi}{5}$$

is the primitive fifth root of unity. Take

$$T_0 = m_0^2 + m_1^2 + m_2^2 + m_3^2 + m_4^2,$$

$$T_1 = m_0 m_1 + m_1 m_2 + m_2 m_3 + m_3 m_4 + m_4 m_0,$$

$$T_2 = m_0 m_2 + m_1 m_3 + m_2 m_4 + m_3 m_0 + m_4 m_1,$$

$$M = m_0 + m_1 + m_2 + m_3 + m_4 = 2018.$$

Therefore, $S = (2 - \lambda)T_1 + (3 + \lambda)T_2$. Note that if we rotate the distribution of points or make symmetry about an axis of symmetry of the regular pentagon, the value of S will not change.

If m_0, m_1, m_2, m_3, m_4 can take real values, then M^2 is an important value when they are all equal. We consider the difference between S and $M^2 = T_0 + 2T_1 + 2T_2$,

$$M^2 - S = T_0 + \lambda T_1 + (-1 - \lambda)T_2 = T_0 + (\omega + \omega^4)T_1 + (\omega^2 + \omega^3)T_2.$$

The above expression can be factorized as

$$M^2 - S$$
$$= (m_0 + m_1 \omega + m_2 \omega^2 + m_3 \omega^3 + m_4 \omega^4)$$
$$\times (m_0 + m_1 \omega^4 + m_2 \omega^3 + m_3 \omega^2 + m_4 \omega).$$

Let complex number $Z = Z(\boldsymbol{m}) = m_0 + m_1 \omega + m_2 \omega^2 + m_3 \omega^3 + m_4 \omega^4$, and we have $M^2 - S = Z\overline{Z} = |Z|^2$. Therefore, the maximization of S is equivalent to the minimization of $|Z|^2$ or $|Z|$.

We consider the parity of the five numbers in $\boldsymbol{m} = (m_0, m_1, m_2, m_3, m_4)$. In any case, there exists an axis of symmetry of the pentagon such that the parity of the number of points on the two vertices symmetric about this axis is the same. After appropriate rotation, it is possible to set m_1 and m_4, m_2 and m_3 to have the same parity. Consider then the following new distribution

$$\boldsymbol{m}^* = \left(m_0, \frac{m_1 + m_4}{2}, \frac{m_2 + m_3}{2}, \frac{m_2 + m_3}{2}, \frac{m_1 + m_4}{2} \right).$$

The five components of $\boldsymbol{m}^*$ are still non-negative integers and their sum is 2018. We will explain that there is $|Z(\boldsymbol{m}^*)| \leq |Z(\boldsymbol{m})|$, and the equal sign holds if and only if $m_1 = m_4, m_2 = m_3$.

In fact,

$$Z(m^*) = \frac{1}{2}(Z(m) + \overline{Z(m)}) = \text{Re}Z(m).$$

Then

$$|Z(m^*)| = |\text{Re}Z(m)| \le |Z(m)|.$$

The equal sign holds if and only if $Z(m)$ is a real number, namely $Z(m) = \overline{Z(m)}$. Then

$$m_0 + m_1\omega + m_2\omega^2 + m_3\omega^3 + m_4\omega^4$$

$$= m_0 + m_4\omega + m_3\omega^2 + m_2\omega^3 + m_1\omega^4.$$

Therefore,

$$(m_1 - m_4) + (m_2 - m_3)\omega + (m_3 - m_2)\omega^2 + (m_4 - m_1)\omega^3 = 0.$$

Since the minimum polynomial of ω over the field of rational numbers is $1 + x + x^2 + x^3 + x^4$, thus $m_1 = m_4, m_2 = m_3$.

Hence, in order to find the minimum value of $|Z|$, we only need to consider the case when $m_1 = m_4$, $m_2 = m_3$. Then we have

$$Z = m_0 + m_1(\omega + \omega^4) + m_2(\omega^2 + \omega^3)$$

$$= m_0 + \lambda m_1 + (-1 - \lambda)m_2 = X - Y\lambda,$$

where $X = m_0 - m_2, Y = m_2 - m_1$. Since 2018 is not divisible by 5, thus X, Y cannot be both zero. λ is an irrational number satisfying equation $x^2 + x - 1 = 0$, and the other root of this equation is $-1 - \lambda$. We introduce the conjugate algebra of Z, namely, $\tilde{Z} = X - Y(-1 - \lambda) = X + Y(1 + \lambda)$.

Then we have $Z \cdot \tilde{Z} = X^2 + XY - Y^2$. Taking $H = X^2 + XY - Y^2 = Z \cdot \tilde{Z}$, hence

$$|Z| = \frac{|H|}{|\tilde{Z}|}.$$

First consider the numerator $|H|$. We have $4H = 4(X^2 + XY - Y^2) = (2X + Y)^2 - 5Y^2$. Since 2, 3 are not square residues of modulus 5, thus $H \ne \pm 2, \pm 3$, namely, $H = 1$ or $|H| \ge 4$.

If $XY \le 0$, since X, Y are not all zero, thus $|Z| = |X - \lambda Y| = |X| + |\lambda Y| \ge \lambda$. In the following we will consider $XY < 0$.

Case 1: $H = \pm 1$. Then there is $|X|^2 + |X| \cdot |Y| - |Y|^2 = H = \pm 1$.
To get positive integer solutions of indefinite equation $X^2 + XY - Y^2 = \pm 1$, let (X, Y) be a positive integer solution. Since $Y^2 = X^2 + XY \mp 1 \geq X^2$, thus $Y \geq X$. If $X = Y$ then $X = Y = 1$. If $X < Y$, since

$$(Y - X)^2 + (Y - X)X - X^2 = -(X^2 + XY - Y^2) = \mp 1,$$

thus $(Y - X, Y)$ is also a positive integer solution. Each solution can be changed to a smaller solution by operation $(X, Y) \to (Y - X, X)$ until reach the minimum solution $(1, 1)$. Vice versa, starting from $(1, 1)$ all the positive integer solutions can be generated using operation $(X, Y) \to (Y, X + Y)$. Therefore, the positive integer solutions (X, Y) of $X^2 + XY - Y^2 = \pm 1$ are the two adjacent terms in the Fibonacci series
$$1, 1, 2, 3, 5, 8, 13, 21, 34, 55, 89, 144, 233, 377, 610, 987, 1597, \ldots, \quad \text{namely,}$$
$(X, Y) = (f_n, f_{n+1})$.

If X, Y are both positive integers, since

$$M = 5m_1 + X + 3Y = 2018,$$

it follows that $X + 3Y \leq 2018$ and $X + 3Y \equiv 2018 \equiv 3 \pmod 5$. Therefore, the maximum value of X is 144 and the corresponding $Y = 233$.
Consequently,

$$\tilde{Z} = X + (1 + \lambda)Y \leq 377 + 233\lambda,$$

$$|Z| = \frac{|H|}{\tilde{Z}} \geq \frac{1}{377 + 233\lambda}.$$

The equal sign is obtained when $X = 144, Y = 233$, corresponding to $m_1 = m_4 = 235$, $m_2 = m_3 = 468$, $m_0 = 612$.
If X, Y are both negative integers, take $X = -U, Y = -V$, and then $(U, V) = (f_n, f_{n+1})$. Since

$$M = 5m_0 + 4U + 2V = 2018,$$

we have $4U + 2V \leq 2018$ and $4U + 2V \equiv 2018 \equiv 3 \pmod 5$. The maximum value of U is 55 and the corresponding $V = 89$. Consequently,

$$|\tilde{Z}| = |U + (1 + \lambda)V| \leq 55 + (1 + \lambda)89 = 114 + 89\lambda,$$

$$|Z| = \frac{|H|}{|\tilde{Z}|} \geq \frac{1}{144 + 89\lambda} > \frac{1}{377 + 233\lambda}.$$

Case 2: $|H| \geq 4$. If X, Y are both positive integers, then

$$|Z| = |X - \lambda Y| = \frac{|H|}{X + (1 + \lambda)Y} \geq \frac{4}{X + 3Y} \geq \frac{4}{2018} > \frac{1}{377 + 233\lambda}.$$

If X, Y are both negative integers, take $X = -U, Y = -V$, and then

$$|Z| = |U - \lambda V| = \frac{|H|}{U + (1 + \lambda)V} \geq \frac{4}{4U + 2V} \geq \frac{4}{2018} > \frac{1}{377 + 233\lambda}.$$

To sum up, $|Z|$ takes its minimum value if and only if $Z = 144 - 233\lambda, m_0 = 612, m_1 = m_4 = 235, m_2 = m_3 = 468$. Therefore, the method of placing 2018 points that maximize S is to start from a vertex and place $612, 235, 468, 468, 235$ points on the five vertices of the regular pentagon in counterclockwise order, respectively. $\square$

China Mathematical Olympiad

2019 (Wuhan, Hubei)

2019 China Mathematical Olympiad (also named the 35th National Mathematics Winter Camp for Middle School Students), was sponsored by the China Mathematical Society and hosted by the Hubei Mathematical Society and the First Affiliated Middle School of Huazhong Normal University. It was held at the First Affiliated Middle School of Huazhong Normal University in Chengdu from November 24 to 30, 2019, with 35 teams from 31 provinces, municipalities and autonomous regions in China's mainland, Hong Kong Special Administrative Region of China, Macao Special Administrative Region of China, Russia and Singapore, including 415 participants.

Two tests were held during the competition, and according to their test scores, 60 students were selected into the national training team. There are 379 senior students from China Mainland participating the competition, among which 132 ones won the gold medals, 155 ones won the silver medals, and 92 ones won the bronze medals.

The members of the main examination committee of this winter camp are:

Xiong Bin (East China Normal University);

Yu Hongbing (Suzhou University);

Yao Yijun (Fudan University);

Qu Zhenhua (East China Normal University);

Liu Ruochuan (Peking University);

Ai Yinghua (Tsinghua University);

He Yijie (East China Normal University);

Wang Bin (Chinese Academy of Sciences Academy of Mathematics and Systems Science);
Wang Xinmao (University of Science and Technology of China);
Luo Ye (Hong Kong University);
Li Weigu (Peking University);
Li Ting (Sichuan University);
Fu Yunhao (Southern University of Science and Technology);
Ji Chungang (Nanjing Normal University);
Li Ming (Nankai University);
Zhang Sihui (University of Shanghai for Science and Technology);
Wang guozhen (Fudan University);
Jin Long (Tsinghua University);
Lai Li (Tsinghua University);
Leng Fusheng (Chinese Academy of Sciences Academy of Mathematics and Systems Science);
Wu Yuchi (East China Normal University);
Lin Tianqi (East China Normal University);
Luo Zhenhua (East China Normal University).

First Day
(8:00 – 12:00; November 26, 2019)

1 Suppose real numbers $a_1, a_2, \ldots, a_{40}$ satisfy $a_1 + a_2 + \cdots + a_{40} = 0$ and $|a_i - a_{i+1}| \leq 1$, for $1 \leq i \leq 40$, where $a_{41} = a_1$. Let $a = a_{10}, b = a_{20}, c = a_{30}, d = a_{40}$.

(1) Find the maximum value of $a + b + c + d$;

(2) find the maximum value of $ab + cd$. (Contributed by He Yijie)

Solution (1) We make the convention that all the subscripts here are under modulus 40. By the given condition, for any integers k, m, we have $a_{m+k} \geq a_m - |k|$. Thus

$$0 = \sum_{i=1}^{40} a_i = \sum_{k=-4}^{5} a_{10+k} + \sum_{k=-4}^{5} a_{20+k} + \sum_{k=-4}^{5} a_{30+k} + \sum_{k=-4}^{5} a_{40+k}$$

$$\geq \sum_{k=-4}^{5} (a - |k|) + \sum_{k=-4}^{5} (b - |k|) + \sum_{k=-4}^{5} (c - |k|) + \sum_{k=-4}^{5} (d - |k|)$$

$$= 10(a + b + c + d) - 4 \times 25,$$

then it follows that $a + b + c + d \leq 10$.

On the other hand, let

$$a_{10i+k} = \frac{5}{2} - |k| (0 \le i \le 3, -4 \le k \le 5).$$

Then $a_1, a_2, \ldots, a_{40}$ satisfy the condition and $a + b + c + d = 4 \times \frac{5}{2} = 10$.
Hence the maximum value of $a + b + c + d$ is 10.

(2) By symmetry, we may assume $ab \ge cd$. If $ab \le 0$, then $ab + cd \le 0$.
In the following we will only consider the case $ab > 0$.

We can also suppose $a, b > 0$ (Otherwise, if $a, b < 0$, we consider $-a_1, -a_2, \ldots, -a_{40}$ instead of $a_1, a_2, \ldots, a_{40}$. This will not affect the condition and does not change the value of $ab + cd$). By the given condition we have

$$0 = \sum_{i=1}^{40} a_i = \sum_{k=-14}^{5} a_{10+k} + \sum_{k=-4}^{15} a_{20+k}$$

$$\ge \sum_{k=-14}^{5} (a - |k|) + \sum_{k=-4}^{15} (b - |k|)$$

$$= (20a - 120) + (20b - 130)$$

$$= 20(a + b) - 250,$$

and then $a + b \le \dfrac{25}{2}$.

(i) When $cd > 0$, if $a + b > 10$, from the conclusion of (1) we have
$c + d < 0$, and thus $c, d < 0$. Since $c \ge b - 10, d \ge a - 10$, so $cd \le (10 - b)(10 - a)$.
Note that $10 < a + b \le \dfrac{25}{2}$, so we have

$$ab + cd \le ab + (10 - b)(10 - a)$$

$$= 2(a - 5)(b - 5) + 50$$

$$\le \frac{1}{2}(a + b - 10)^2 + 50$$

$$\le \frac{425}{8}.$$

If $a + b \le 10$, then

$$ab + cd \le 2ab \le \frac{(a + b)^2}{2} \le 50 < \frac{425}{8}.$$

(ii) When $cd \leq 0$, we have

$$ab + cd \leq ab \leq \frac{(a+b)^2}{4} \leq \frac{625}{16} < \frac{425}{8}.$$

On one hand, from (i), (ii) we get $ab + cd \leq \dfrac{425}{8}$.

On the other hand, let

$$a_{10+k} = \frac{25}{4} - |k|(-14 \leq k \leq 5), \quad a_{20+k} = \frac{25}{4} - |k|(-4 \leq k \leq 15).$$

Then $a_1, a_2, \ldots, a_{40}$ satisfy the condition, and

$$ab + cd = \left(\frac{25}{4}\right)^2 + \left(-\frac{15}{4}\right)^2 = \frac{425}{8}.$$

Consequently, the maximum value of $ab + cd$ is $\dfrac{425}{8}$.   □

2   As shown in Fig. 2.1, in $\triangle ABC$, $AB > AC$, the bisector of $\angle BAC$ intersects with side BC at point D. Point P lies on the extension of line DA, and PQ is tangent to the circumscribed circle of $\triangle ABD$ at point Q (Q, B are on the same side of line AD). PR is tangent to the circumscribed circle of $\triangle ACD$ at point R (R, C are on the same side of line AD). Line segments BR, CQ intersect at point K. Make parallel line of BC through K and intersects with QD, AD, RD at the points E, L, F, respectively. Prove that $EL = FK$. (Contributed by Xiong Bin)

Solution   Suppose the circumcenters of $\triangle ABD$ and $\triangle ACD$ are O_1, O_2, with radii r_1, r_2, respectively. The distances from points O_1, O_2 to line BC

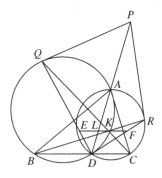

Fig. 2.1

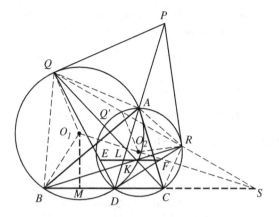

Fig. 2.2

are d_1, d_2, respectively. Extend O_1O_2 to intersect with the extension of BC at point S.

Construct $O_1M \perp BD$ at point M, so

$$d_1 = O_1M = BO_1 \cos \angle MO_1B$$

$$= r_1 \cos \angle BAD.$$

Similarly, we obtain $d_2 = r_2 \cos \angle CAD$.

Combing $\angle BAD = \angle CAD$, we have $\dfrac{d_1}{d_2} = \dfrac{r_1}{r_2}$, namely, $\dfrac{SO_1}{SO_2} = \dfrac{r_1}{r_2}$.
Therefore, $\odot O_1, \odot O_2$ are homothetic with respect to S.

Suppose line segment QR intersects with circle $\odot O_2$ at another point Q'. By the circle power theorem,

$$PQ^2 = PA \cdot PD = PR^2,$$

and then $PQ = PR$. Therefore,

$$\angle O_2Q'R = \angle O_2RQ'$$

$$= 90° - \angle PRQ$$

$$= 90° - \angle PQR$$

$$= \angle O_1QR,$$

and then $O_1Q // Q_2Q'$.

Since $\dfrac{O_1Q}{O_2Q'} = \dfrac{r_1}{r_2} = \dfrac{O_1S}{O_2S}$, then Q, Q' are the homothetic corresponding points of $\odot O_1, \odot O_2$, and hence S, Q, Q' are collinear, namely, S, Q, R are collinear.

It should be noted that D, C are also the homothetic corresponding points of $\odot O_1, \odot O_2$. Combing the circle power theorem, we have

$$SQ \cdot SR = \frac{r_1}{r_2} SQ' \cdot SR = \frac{r_1}{r_2} SC \cdot SD = SD^2,$$

and thus $\triangle SQD \sim \triangle SDR$. Then

$$\angle DFE = \angle RDS = \angle DQS = \angle DQR,$$

and thus $\triangle DEF \sim \triangle DRQ$. By the laws of sines on area, it follows that

$$\frac{EL}{FL} = \frac{S_{\triangle DEL}}{S_{\triangle DFL}} = \frac{DE}{DF} \cdot \frac{\sin \angle EDL}{\sin \angle FDL}$$

$$= \frac{DR}{DQ} \cdot \frac{\sin \angle QDA}{\sin \angle RDA}$$

$$= \frac{DR}{DQ} \cdot \frac{\dfrac{AQ}{2r_1}}{\dfrac{AR}{2r_2}}$$

$$= \frac{DR}{DQ} \cdot \frac{AQ}{AR} \cdot \frac{r_2}{r_1}.$$

Since $\triangle PAQ \sim \triangle PQD, \triangle PAR \sim \triangle PRD$, thus

$$\frac{AQ}{QD} = \frac{PA}{PQ} = \frac{PA}{PR} = \frac{AR}{RD},$$

and then $\dfrac{EL}{FL} = \dfrac{r_2}{r_1}$.

As $\angle BQD = \angle BAD = \angle CAD = \angle CRD$, then

$$\angle BQS = \angle BQD + \angle DQS = \angle CRD + \angle SDR = \angle RCS,$$

and thus B, C, R, Q are concyclic. Therefore, $\triangle KBQ \sim \triangle KCR, \dfrac{QK}{RK} = \dfrac{BQ}{CR}$. Furthermore,

$$\angle BRD = \angle BRC - \angle CRD = \angle BQC - \angle BQD = \angle CQD.$$

Combing the laws of sines, we have

$$\frac{EK}{FK} = \frac{\dfrac{EK}{\sin\angle EQK}}{\dfrac{FK}{\sin\angle FRK}} = \frac{\dfrac{QK}{\sin\angle QEK}}{\dfrac{RK}{\sin\angle RFK}}$$

$$= \frac{QK}{RK} \cdot \frac{\sin\angle RFK}{\sin\angle QEK}$$

$$= \frac{BQ}{CR} \cdot \frac{\sin\angle RDC}{\sin\angle QDB}$$

$$= \frac{r_1}{r_2}.$$

Consequently, $\dfrac{EK}{FK} = \dfrac{r_1}{r_2} = \dfrac{FL}{EL}$, namely $EL = FK$. $\qquad\square$

3. Let S be a set of 35 elements and F be a set consisting of some maps from S to itself. For positive integer k, we say F has property $P(k)$ if for any $x, y \in S$, there exist k mappings $f_1, f_2, \ldots, f_k$ (which can be the same) such that

$$f_k(\cdots(f_2(f_1(x)))\cdots) = f_k(\cdots(f_2(f_1(y)))\cdots).$$

Find the smallest positive integer m such that all F with property $P(2019)$ have property $P(m)$. (Contributed by Wang Xinmao)

Solution The required m is 595.

Let $\mathcal{U}$ be the set consisting of all the two-element subsets of S. We call $V \in \mathcal{U}$ a "good set" if there exists $f \in F$ such that $f(V)$ is a one-element set.

Construct a directed graph G in the following way: Regard $\mathcal{U}$ as the set of vertices; for $U_1, U_2 \in \mathcal{U}$, if there exists $f \in F$ such that $U_2 = f(U_1)$, then a directed edge $U_1 \to U_2$ pointing from U_1 to U_2 is called an f-edge from U_1 to U_2. In this way, "F has property $P(k)$" is equivalent to "for any $U \in \mathcal{U}$, there exists a directed path with length no more than $k-1$, starting from U and ending at some good set".

Firstly, we prove that every F with property $P(2019)$ will have property $P(595)$. Suppose F has property $P(2019)$. Then for any $U \in \mathcal{U}$, there exists a directed path from U to some good set.

We consider the shortest path with this property

$$U = U_0 \to U_1 \to \cdots \to U_m.$$

Then $U_0, \ldots, U_m$ are distinct two-element sets, and thus $m+1 \leq \mathrm{C}_{35}^2 = 595$, i.e., there exists a path from U to a good set with length not exceeding 594. This proves that F has property $P(595)$.

Next, we construct a set F such that it has property $P(595)$ but not property $P(594)$. It follows that F has property $P(2019)$, and that F does not have property $P(m)$ for any positive integer $m \leq 594$.

It may be appropriate to set $S = \{1, 2, \ldots, 35\}$. For $U = \{x, y\} \subset S$, we define

$$d(U) = \min(|x - y|, 35 - |x - y|).$$

For a directed edge $U \to V$, if $d(V) - d(U) = 0, 1, -1$, then this edge is said to be a distance-preserving, distance+1 and distance−1 edge, respectively. We define $f_1, f_2 : S \to S$ as

$$f_1(x) = \begin{cases} x + 1, & 1 \leq x \leq 34, \\ 1, & x = 35, \end{cases}$$

$$f_2(x) = \begin{cases} x, & 1 \leq x \leq 34, \\ 1, & x = 35. \end{cases}$$

It is easy to verify that $F = \{f_1, f_2\}$ has the following properties:

(1) $\{1, 35\}$ is the only good set of F.
(2) The f_1 edges are all distance-preserving. There are 33 f_2 edges that are not cyclic edges; 16 of which are distance−1 edges ($\{i, 35\} \to \{1, i\}(2 \leq i \leq 17)$), 16 are distance+1 edges ($\{i, 35\} \to \{1, i\}(19 \leq i \leq 34)$), and 1 is a distance-preserving edge ($\{18, 35\} \to \{18, 1\}$).

Consider the shortest path L from $\{1, 18\}$ to the good set $\{1, 35\}$:

$$\{1, 18\} = U_0 \to U_1 \to \cdots \to U_m = \{1, 35\}.$$

Notice that $d(U_0) = 17, d(U_m) = 1$, so the distance decreases by 16. By property (2), L contains every distance−1 f_2 edge exactly once, and does not contain any distance+1 or distance-preserving f_2 edge (because the distance-preserving f_2 edge ends at U_0). Moreover, all the edges in L are f_1 edges. Therefore, L is as follows:

$$U_0 \xrightarrow{\ 34\ f_1\ \textbf{edges}\ } \{17,\ 35\} \xrightarrow{\quad f_2 \quad} \{1,\ 17\} \xrightarrow{\ 34\ f_1\ \textbf{edges}\ } \{16,\ 35\}$$

$$\xrightarrow{\quad f_2 \quad} \{1,\ 16\} \to \cdots \to \{1,\ 2\} \xrightarrow{\ 34\ f_1\ \textbf{edges}\ } \{1,\ 35\}.$$

Therefore, the length of L is $35 \times 16 + 34 = 594$. This indicates that the shortest distance from U_0 to the good set is 594. Notice that this path traverses every member of $\mathcal{U}$, so F has property $P(595)$ but not property $P(594)$. $\qquad\qquad\qquad\qquad\qquad\qquad\qquad\qquad\qquad\qquad\qquad\qquad\quad$ □

Second Day
(8:00 – 12:30; November 27, 2019)

4 Find the largest real number c such that the following conclusion holds for any integer $n \geq 3$: Let $A_1, A_2, \ldots, A_n$ be n arcs on the circumference of a circle (each arc contains its own endpoints). If there exist at least $\frac{1}{2}C_n^3$ triples (i, j, k) satisfying $1 \leq i < j < k \leq n$ and $A_i \cap A_j \cap A_k \neq \varnothing$, then there exists $I \subset \{1, 2, \ldots, n\}$ satisfying $|I| > cn$ and $\underset{i \in I}{\cap} A_i \neq \varnothing$. (Contributed by Ai Yinghua)

Solution The largest real number c required is $\dfrac{\sqrt{6}}{6}$.

At first, we will prove that the conclusion of the question holds for $c = \dfrac{\sqrt{6}}{6}$.

Let the circumference of the given circle be oriented in a clockwise direction, so that the startpoint and endpoint of an arc are determined in this direction. Notice that if $A_i \cap A_j \neq \varnothing$, then either the startpoint of A_i is contained in A_j or that of A_j is contained in A_i, or both are true. Construct a directed graph G with vertex set $\{A_1, A_2, \ldots, A_n\}$: For $i \neq j$, if the startpoint of A_i is in A_j, then connect directed edge $A_j \to A_i$.

For any set $\{A_q \to A_p, A_r \to A_p\}$ $(q \neq r)$ that contains two directed edges with the same end vertex, let p, q, r be arranged from smallest to largest as $i < j < k$, then $A_i \cap A_j \cap A_k = A_p \cap A_q \cap A_r$ contains the startpoint of A_p and is non-empty. On the other hand, if $A_i \cap A_j \cap A_k \neq \varnothing, i < j < k$, take any point X in the intersection, and, starting from it, move counterclockwise along the circumference, to meet the startpoint of A_p in A_i, A_j, A_k at first (If there is another arc with the same startpoint, then take any one of them). Denoting $\{q, r\} = \{i, j, k\} \backslash \{p\}$, then p, q, r are the permutations of i, j, k, and $A_q \to A_p, A_r \to A_p$. This shows that the mapping Φ from set S (consisting of sets each containing two directed edges $A_q \to A_p, A_r \to A_p (q \neq r)$) to set T (consisting of triples (i, j, k) satisfying the conditions in the question) is an onto mapping. Here

$$\Phi(\{A_q \to A_p, A_r \to A_p\}) = (i, j, k),$$

where $\{i, j, k\} = \{p, q, r\}$ and $i < j < k$.

Denoting the in-degree of A_i in graph G as $d_-(A_i)$, then

$$|S| = \sum_{i=1}^{n} C_{d_-(A_i)}^2 \geq |T| \geq \frac{1}{2} C_n^3.$$

By the mean value theorem, there exists $1 \leq k \leq n$ satisfying

$$C_{d_-(A_k)}^2 \geq \frac{1}{2n} C_n^3 = \frac{1}{12}(n-1)(n-2).$$

Equivalently,

$$d_-(A_k) \geq \frac{1}{2} + \sqrt{\frac{1}{4} + \frac{(n-1)(n-2)}{6}}.$$

Let $I = i \mid A_i \to A_k \cup \{k\}$. For any $j \in I$, A_j contains the startpoint of A_k. Therefore, $\underset{j \in I}{\cap} A_j \neq \varnothing$ and

$$|I| = d_{-1}(A_k) + 1$$
$$\geq \frac{3}{2} + \sqrt{\frac{1}{4} + \frac{(n-1)(n-2)}{6}}$$
$$> \frac{3}{2} + \frac{n-2}{\sqrt{6}}$$
$$> \frac{n}{\sqrt{6}}.$$

This proves that $c = \dfrac{\sqrt{6}}{6}$ satisfies the condition.

To prove $c \leq \dfrac{\sqrt{6}}{6}$, consider the following construction: for large integer n, let

$$a = \left\lceil \frac{1}{2} + \sqrt{\frac{1}{4} + \frac{(n-1)(n-2)}{6}} \right\rceil < \frac{n}{2}.$$

Take n equally distributed points on the circumference of the circle and record them in a clockwise order as $P_1, P_2, \ldots, P_n$. Let A_i be the minor arc $\widehat{P_i P_{i+a}}$, with startpoint P_i and endpoint P_{i+a} (the subscript is under modulo n). Since the startpoint of each arc is different from each other, and the union of any two arcs is not the whole circumference, then $A_q \to A_p$ and

$A_p \to A_q$ cannot be valid at the same time. Therefore, mapping $\Phi : S \to T$ defined before is injective, so $|T| = |S|$.

Note also that each A_i in G has exactly a incoming edges, i.e. $d_-(A_i) = a$, so we haves

$$|T| = |S| = \sum_{i=1}^{n} C_{d_- A_i}^2 = n C_a^2 \geq \frac{1}{2} C_n^3.$$

If $I \subset \{1, 2, \ldots, n\}$ satisfies $\underset{i \in I}{\cap} A_i \neq \emptyset$, take a point X in the intersection and, starting from it, move along the circumference in counterclockwise direction, until meet firstly the startpoint of $A_i (i \in I)$, which will be regarded as the starting point of $A_p (p \in I)$. Then for any $q \in I, q \neq p$, there is $A_q \to A_p$. Therefore,

$$cn < |I| \leq d_{-1}(A_p) + 1 = a + 1.$$

Consequently,

$$c \leq \lim_{n \to \infty} \frac{1 + \left[\frac{1}{2} + \sqrt{\frac{1}{4} + \frac{(n-1)(n-2)}{6}} \right]}{n} = \frac{\sqrt{6}}{6}.$$

In summary, the largest real number c required is $\dfrac{\sqrt{6}}{6}$. $\qquad\square$

5 Sequence $\{a_n\}_{n \geq 1}$ is defined as follows: a_1 is an integer greater than 1. For $n \geq 1$, there is $a_{n+1} = a_n + P(a_n)$, where $P(a_n)$ denotes the largest prime factor of a_n. Prove that there are perfect squares in sequence $\{a_n\}_{n \geq 1}$. (Contributed by Qu Zhenhua)

Solution 1 Since $P(a_n) \mid a_n + P(a_n) = a_{n+1}$, then $P(a_{n+1}) \geq P(a_n)$.

Claim 1 $\{P(a_n)\}$ *does not have an upper bound.*

If $\{P(a_n)\}$ has an upper bound, then it is eventually a constant, so there exists positive integer N as well as prime number p such that when $n \geq N$, there are always $P(a_n) = p$. Let $a_N = pt$. By the definition of the sequence, we have $a_{N+i} = p(t + i)$ for $i \geq 0$. We can take a positive integer i such that $t + i = q$ is a prime number greater than p, and then $P(a_{N+i}) = P(pq) = q > p$. This is a contradiction. Therefore, $\{P(a_n)\}$ does not have an upper bound.

Let $I = i \geq 2 \mid P(a_i) > P(a_{i-1})$. From Claim 1 we know that I is an infinite set. We denote the elements in I from smallest to largest as

$$i_2 < i_3 < i_4 < \cdots$$

and let $i_1 = 1$. For each $j \geq 1$, let $P(a_{i_j}) = p_j$, and then $p_1 < p_2 < p_3 < \cdots$. For $j \geq 2$. Since

$$P(a_{i_j-1}) = p_{j-1}, a_{i_j} = a_{i_j-1} + p_{j-1},$$

then $p_{j-1} \mid a_{i_j}$. As $p_j \mid a_{i_j}$, then

$$a_{i_j} = p_{j-1} p_j m_j, j \geq 2,$$

where m_i is positive integer.

Claim 2 $m_{j\,j\geq 2}$ *is monotonic decreasing.*

Let $j \geq 2$. Since $P(a_k) = p_j$ for $i_j \leq k < i_{j+1}$, then

$$a_k = a_{i_j} + (k - i_j)p_j = p_j(p_{j-1}\, m_j + k - i_j), i_j \leq k \leq i_{j+1}.$$

Since $p_{j-1}m_j, p_{j-1}\, m_j + 1, \ldots, p_{j-1}\, m_j + (i_{j+1} - i_j - 1)$ are not divisible by p_{j+1}, then $i_{j+1} - i_j < p_{j+1}$. Consequently,

$$m_{j+1} = \frac{a_{i_{j+1}}}{p_j p_{j+1}} = \frac{a_{i_j} + (i_{j+1} - i_j)p_j}{p_j p_{j+1}}$$

$$< \frac{p_{j-1}p_j\, m_j + p_{j+1}p_j}{p_j p_{j+1}} < m_j + 1,$$

namely, $m_{j+1} \leq m_j$.

Since m is a monotonic decreasing sequence of positive integers, then it is ultimately a constant. Therefore, there exist positive integers $j_0 \geq 2, m$, such that $m_j = m$ when $j \geq j_0$.

Claim 3 $m = 1$.

If $m \geq 2$, take a sufficiently large prime q such that $q > mp_{j_0}$. Then there exists $j \geq j_0$ such that $mp_j < q < mp_{j+1}$. For $i_j \leq k \leq i_{j+1}$, we have

$$a_k = a_{i_j} + (k - i_j)p_j = p_j(p_{j-1}m + k - i_j),$$

while $a_{i_{j+1}} = p_j p_{j+1} m$. The prime factors of numbers $p_{j-1}m, p_{j-1}m + 1, \ldots, p_{j+1}m - 1$ do not exceed p_j, but

$$q \in p_{j-1}m, p_{j-1}\, m+1, \ldots, p_{j+1}\, m - 1,$$

where $q > mp_j > p_j$. This is a contradiction. Therefore, $m = 1$.

For $j \geq j_0$, there is $a_{i_j} = p_{j-1}\ p_j$. By the definition of the sequence, we have

$$a_{i_j + p_j - p_j - 1} = p_{j-1}\ p_j + (p_j - p_{j-1})p_j = p_j^2.$$

This is a perfect square. Then we finish the proof. $\qquad\square$

Solution 2 Let $p_n = P(a_n), a_n = p_n b_n$. If there exists positive integer n such that $b_n \leq p_n$, then by the definition of the given sequence we have

$$
\begin{aligned}
a_{n+p_n-b_n} &= a_n + p_n(p_n - b_n) \\
&= p_n b_n + p_n(p_n - b_n) \\
&= p_n^2.
\end{aligned}
$$

Therefore, the conclusion holds. In the following we suppose that for any n there is $b_n > p_n$.

We will show that $b_{n+1} \leq b_n + 1$. In fact, there is

$$a_{n+1} = p_n b_n + p_n, p_n \mid a_{n+1},$$

and thus $p_{n+1} \geq p_n$. Consequently,

$$b_{n+1} = \frac{p_n b_n + 1}{p_{n+1}} \leq b_n + 1.$$

Since a_n does not have an upper bound and $b_n \geq \sqrt{a_n}$, then b_n also has no upper bound. Let p be a prime number greater than b_1. Then there exists m such that $b_m > p > b_1$. As $b_{n+1} \leq b_n + 1$, by the discrete version of intermediate value theorem, it follows that there exists positive integer k such that $b_k = p$. Thus we have $a_k = p_k b_k = p_k p$. But $p = b_k > p_k$, an contradiction to the definition of p_k. Therefore, there always exists n such that $b_n \leq p_n$. Consequently, the conclusion holds. $\qquad\square$

6 Do there exist positive real numbers $a_0, a_1, \ldots, a_{19}$ satisfying the following two conditions?

(i) Polynomial $P(x) = x^{20} + a_{19}x^{19} + \cdots + a_1 x + a_0$ has no real roots.

(ii) For any integers $0 \leq i < j \leq 19$, the polynomial obtained by exchanging the coefficients of x^i and x^j in $P(x)$ has real roots.

Please prove your conclusion. (Contributed by Qu Zhenhua)

Solution It does exist. Consider the polynomial

$$P_t(x) = x^{20} + (t+1)x^{19} + (t+20)x^{18} + (t+2)x^{17}$$
$$+(t+19)x^{16} + \cdots + (t+10)x + (t+11).$$

The coefficients of $x^0, x^2, \ldots, x^{18}$ are $t+11, t+12, t+13, \ldots, t+20$, respectively, and that of $x^{19}, x^{17}, \ldots, x^1$ are $t+1, t+2, \ldots, t+10$, respectively, where $t \geq 0$ is a parameter to be determined.

(1) $P_0(x)$ has no real roots.

Simply denoting $P_0(x) = x^{20} + c_{19}x^{19} + \cdots + c_0$, then we have

$$\min\{c_{18}, c_{16}, \ldots, c_0\} > \max\{c_{19}, c_{17}, \ldots, c_1\}.$$

By the inequality of arithmetic and geometric means, for any real number x and $i = 1, 3, 5, \ldots, 17$, there is

$$\frac{c_{i-1}}{2}x^{i-1} + \frac{c_{i+1}}{2}x^{i+1} \geq \sqrt{c_{i-1}\,c_{i+1}}|x|^i \geq c_i|x|^i,$$

and also $x^{20} + \frac{c_{18}}{2}x^{18} \geq c_{19}|x|^{19}, \frac{c_0}{2} > 0$. Adding the above inequalities together, then

$$P(x) \geq \left(x^{20} + \frac{c_{18}}{2}x^{18} - c_{19}|x|^{19}\right)$$
$$+ \sum_{i=1}^{9}\left(\frac{c_{2i-1}}{2}x^{2i-1} + \frac{c_{2i+1}}{2}x^{2i+1} - c_{2i}|x|^{2i}\right) + \frac{c_0}{2} > 0.$$

(2) There exists $t > 0$ such that $P_t(x)$ has real roots.

Consider polynomial $f(t) = P_t(-\frac{t}{2}) = -\frac{1}{2^{20}}t^{20} + \frac{1}{2^{19}}t^{19} + \cdots + 11$. It is easy to see that $f(t)$ is negative when t is sufficiently large, and $f(0) = P_0(0) = 11 > 0$. Then $f(t)$ has positive real roots, i.e. there exists $t > 0$ such that

$$f(t) = P_t(-\frac{t}{2}) = 0.$$

In the following we take positive real number u such that $P_u(x)$ has real roots. Denote

$$S = \{t \in [0,\ u] \mid P_t(x) \text{ has no real roots}\},$$
$$T = \{t \in [0,\ u] \mid P_t(x) \text{ has real roots}\},$$

and then $0 \in S, u \in T$.

(3) For any t, every real root of $P_t(x)$ lies in interval $\left(-u - 1, -1 - \dfrac{1}{u + 10}\right)$.

Obviously, for $x \geq 0$, we have $P_t(x) > 0$. For $-1 - \dfrac{1}{u + 10} \leq x < 0$, we have

$$P_t(x) = x^{20} + ((t + 1)x^{19} + (t + 20)x^{18}) + \cdots$$
$$+ ((t + 10)x + (t + 11)) > 0.$$

For $x \leq -u - 1$, we have

$$P_t(x) = (x^{20} + (t + 1)x^{19}) + \cdots + ((t + 12)x^2 +$$
$$(t + 10)x) + (t + 11) > 0.$$

Therefore, the real roots of $P_t(x)$ lie in $\left(-u - 1, -1 - \dfrac{1}{u + 10}\right)$.

Take a sufficiently small positive real number ε satisfying

$$\varepsilon((u + 1)^{19} + (u + 1)^{18} + \cdots + (u + 1) + 1) < \frac{1}{u + 10}.$$

Then take a sufficiently large positive integer N satisfying $\dfrac{u}{N} < \varepsilon$. Divide $[0, u]$ into N equal parts by dividing points

$$0, \frac{u}{N}, \frac{2u}{N}, \ldots, \frac{N - 1}{N}u, u.$$

Since $0 \in S, u \in T$, there exist two adjacent dividing points $\dfrac{ku}{N} \in S, \dfrac{k + 1}{N}u \in T$, w$0 \leq k \leq N - 1$.

(4) $P(x) = P_{\frac{ku}{N}}(x)$ satisfies the condition.

Denote $Q(x) = P_{\frac{k+1}{N}u}(x)$. As $\dfrac{ku}{N} \in S$, then $P(x)$ has no real roots. Condition (i) is then satisfied. As $\dfrac{k + 1}{N}u \in T$, $Q(x)$ has real roots. Suppose a is a real root of $Q(x)$, from (3) we learn that

$$-u - 1 < a < -1 - \frac{1}{u + 10}.$$

Since $\dfrac{u}{N} < \varepsilon$, by the choice of ε, it follows that

$$|P(a) - Q(a)| < \varepsilon(|a|^{19} + |a|^{18} + \cdots + |a| + 1)$$

$$< \frac{1}{u + 10}.$$

Since $P(x)$ has no real roots, then $0 < P(a) < \dfrac{1}{u + 10}$.
For simplicity, let $P(x) = x^{20} + b_{19}x^{19} + \cdots + b_0$.

The polynomial obtained by exchanging the coefficients b_i, b_j of x^i, x^j in $P(x)$ is denoted by $R(x)$. We will show that $R(a) < 0$, and thus $R(x)$ has real roots because $R(x)$ is positive when x is sufficiently large. It is easy to get

$$P(a) - R(a) = b_i a^i + b_j a^j - b_j a^i - b_i a^j$$

$$= a^j(a^{i-j} - 1)(b_i - b_j)$$

$$= a^i(a^{j-i} - 1)(b_j - b_i).$$

Case 1: $2|i, 2|j$. Then $b_j - b_i \geq 1, a^i \geq 1, a^{j-i} - 1 \geq \dfrac{1}{u + 10}$. Therefore,

$$P(a) - R(a) \geq \frac{1}{u + 10},$$

and hence $R(a) < 0$.

Case 2: $2 \nmid i, 2 \nmid j$. Then $b_j - b_i \leq -1, a^i \leq -1, a^{j-i} - 1 \geq \dfrac{1}{u + 10}$. Therefore,

$$P(a) - R(a) \geq \frac{1}{u + 10},$$

and hence $R(a) < 0$.

Case 3: $2 \nmid i, 2 \mid j$. Then $b_i - b_j \leq -1, a^j \geq 1, a^{i-j} - 1 \leq -1$. Therefore,

$$P(a) - R(a) \geq 1,$$

and hence $R(a) < 0$.

Case 4: $2 \mid i, 2 \nmid j$. Then $b_i - b_j \geq 1, a^j \leq -1, a^{i-j} - 1 \leq -1$. Therefore,

$$P(a) - R(a) \geq 1,$$

and hence $R(a) < 0$.

To sum up, $P(x)$ satisfies the given conditions. $\qquad\square$

China National Team
Selection Test

The first stage of training and selection of the 2019 China National Team for the 60ᵗʰ IMO, was held from March 2 to 10, 2019 at the Middle School Affiliated to South China Normal University in Guangdong Province. The main task was to select Chinese national team members to participate in the 60th International Mathematical Olympiad in United Kingdom in 2019. 60 players participated in the first stage training. After two tests (with equal weights), 19 players were selected to enter the second stage.

The second stage of training and selection was held at the Shanghai High School, Shanghai, from March 20 to 28, 2019. During this period, another two tests were conducted. Finally, according to the total scores of the four tests (with equal weights), The top six scorers were identified to be the Chinese national team members for the 60th IMO. They are Deng Mingyang (The Middle School Affiliated to Renmin University of China, first grade student), Hu Sulin (The Middle School Affiliated to South China Normal University, second grade student), Xie Baiting (Zhilin Middle School in Zhejiang Province, third grade student), Huang Jiajun (Shanghai High School in Shanghai, first grade student), Yuan Zhizhen (Wugang No.3 Middle School in Hubei Province, second grade student), and Yu Ranfeng (The Middle School Affiliated to Nanjing Normal University in Jiangsu Province, second grade student).

The coaches of the national training team are:
Xiong Bin (East China Normal University);
Wu Jianping (Capital Normal University);
Qu Zhenhua (East China Normal University);
Yu Hongbing (Suzhou University);
Leng Gangsong (Shanghai University);
Yao Yijun (Fudan University);
Ai Yinghua (Tsinghua University);
Wang Bin (Institute of Mathematics and Systems Science, Chinese Academy of Sciences);
Fu Yunhao (Guangdong Second Normal University);
He Yijie (East China Normal University);
Zhang Sihui (University of Shanghai for Science and Technology);
Lin Tianqi (East China Normal University).

Test I, First Day
(8:00 – 12:30; March 4, 2019)

1 As shown in Fig. 1.1, convex pentagon $ABCDE$ is inscribed in $\odot O$ and $AB = CD = EA$. Diagonals BD, CE intersect at point P. Point H is the orthocenter of $\triangle ABC$ and M, N are the midpoints of BC, DE, respectively. Point G is the center of gravity of $\triangle AMN$ and lines PH, OG intersect at point T. Prove that $AT \perp CD$. (Contributed by Lin Tianqi)

Solution 1 As shown in Fig. 1.2, let H_1 be the orthocenter of $\triangle ACD$. We only need to prove that points A, T, H_1 are collinear. By the fact

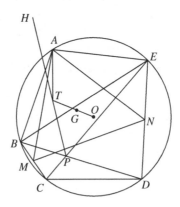

Fig. 1.1

that $\triangle ABE$ is an isosceles triangle, we know that the points O, A, H are collinear. Extend CH_1, DH_1 to intersect with $\odot O$ at points B', E', respectively.

Since $AB = CD$, then $AD//BC$. As $CB'\perp AD$, then $BC\perp B'C$, and thus BB' is the diameter of $\odot O$. Similarly, EE' is also the diameter of $\odot O$. Applying Pascal's theorem to generalized hexagon $B'BDE'EC$ inscribed in the circle, we find that points O, P, H_1 are collinear.

Take the midpoints U, V of chords BE, CD. It is known that

$$\overrightarrow{OU} = \frac{1}{2}\overrightarrow{AH}, \overrightarrow{OV} = \frac{1}{2}\overrightarrow{AH_1}, \quad \text{and then} \quad \overrightarrow{UV} = \frac{1}{2}\overrightarrow{HH_1}.$$

Therefore, $HH_1//UV$.

Since $AB = CD = EA$, then $AD//BC, AC//DE$. Thus $S_{\triangle ABD} = S_{\triangle ACD} = S_{\triangle ACE}$, and then $S_{\triangle APB} + S_{\triangle APD} = S_{\triangle APC} + S_{\triangle APE}$. Since U, V are the midpoints of BE, CD, there is

$$S_{\triangle APU} = \frac{1}{2}(S_{\triangle APE} - S_{\triangle APB}) = \frac{1}{2}(S_{\triangle APD} - S_{\triangle APC}) = S_{\triangle APV},$$

and thus $AP//UV$. Therefore, $AP//HH_1$.

Since G is the center of gravity of $\triangle AMN$, then

$$3\overrightarrow{OG} = \overrightarrow{OA} + \overrightarrow{OM} + \overrightarrow{ON} = \overrightarrow{OA} + \frac{\overrightarrow{OB} + \overrightarrow{OC}}{2} + \frac{\overrightarrow{OD} + \overrightarrow{OE}}{2}$$

$$= \frac{(\overrightarrow{OA} + \overrightarrow{OB} + \overrightarrow{OE}) + (\overrightarrow{OA} + \overrightarrow{OC} + \overrightarrow{OD})}{2} = \frac{\overrightarrow{OH} + \overrightarrow{OH_1}}{2}.$$

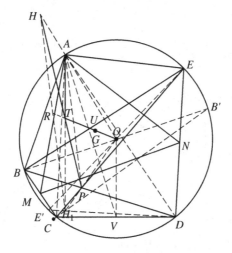

Fig. 1.2

Thus, the intersection R of OG, HH_1 is the midpoint of HH_1. Combing $AP//HH_1$, in $\triangle OHH_1$ there is

$$\frac{OA}{AH} \cdot \frac{HR}{RH_1} \cdot \frac{H_1P}{PO} = \frac{OA}{AH} \cdot 1 \cdot \frac{H_1P}{PO} = 1.$$

By the converse of Ceva theorem, it follows that AH_1, OG, HH_1 are concurrent, and the intersection is point T. Therefore, $AT \perp CD$. $\square$

Solution 2   The complex plane is established with O as the origin. Let the complex numbers corresponding to points $A, B, C, \ldots$ be $a, b, c, \ldots$ and so on. Let $a = 1$, so the equation of $\odot O$ is $|z| = 1$. Suppose the midpoint of is X. Then there exists unit complex number ω such that $c = x\omega^{-1}$, $d = x\omega$. Since $AB = AE = CD$, then $b = \omega^2, e = \omega^{-2}$. By the properties of orthocenter and center of gravity, we have

$$h = a + b + e = 1 + \omega^2 + \omega^{-2},$$

$$g = \frac{1}{3}(a + m + n) = \frac{1}{3}\left(a + \frac{b+c}{2} + \frac{d+e}{2}\right)$$

$$= \frac{2 + \omega^2 + x\omega^{-1} + x\omega + \omega^{-2}}{6} = \frac{(\omega + \omega^{-1})(x + \omega + \omega^{-1})}{6}.$$

Suppose $u = \omega + \omega^{-1} = 2\mathrm{Re}\,\omega \in \mathbb{R}$, and then $h = u^2 - 1 \in \mathbb{R}$, $g = \dfrac{u(x + u)}{6}$.

Since P is the intersection of BD, CE, then $\dfrac{p-b}{d-b} \in \mathbb{R}$ and $\dfrac{p-c}{d-c} \in \mathbb{R}$, namely,

$$\begin{cases} (\bar{d} - \bar{b})(p - b) - (d - b)(\bar{p} - \bar{b}) = 0, \\ (\bar{e} - \bar{c})(p - c) - (e - c)(\bar{p} - \bar{c}) = 0. \end{cases}$$

Note that $\bar{b} = b^{-1}, \bar{c} = c^{-1}, \bar{d} = d^{-1}, \bar{e} = e^{-1}$. Therefore,

$$\begin{cases} (bd)^{-1}(p - b) + (\bar{p} - b^{-1}) = 0, \\ (ce)^{-1}(p - c) + (\bar{p} - c^{-1}) = 0, \end{cases}$$

The solution is

$$p = \frac{b^{-1} + d^{-1} - c^{-1} - e^{-1}}{(bd)^{-1} - (ce)^{-1}} = \frac{\omega^{-2} + x^{-1}\omega^{-1} - x^{-1}\omega - \omega^2}{x^{-1}\omega^{-3} - x^{-1}\omega^3}$$

$$= \frac{1 + x(\omega + \omega^{-1})}{1 + \omega^2 + \omega^{-2}} = \frac{1 + xu}{h}.$$

Since T is the intersection of PH, OG, then $\dfrac{t-h}{h-p} \in \mathbb{R}$, $\dfrac{t}{g} \in \mathbb{R}$. Therefore,

$$\begin{cases} (\bar{h}-\bar{p})(t-h)-(h-p)(\bar{t}-\bar{h})=0, \\ \bar{g} \cdot t - g \cdot \bar{t} = 0. \end{cases}$$

The solution is

$$t = \frac{\bar{h} \cdot p - h \cdot \bar{p}}{\bar{h}-\bar{p}-g^{-1}\bar{g}(h-p)}.$$

Since O, A, H are collinear, h is a real number. Then $\bar{h}=h$, $\bar{p}=\dfrac{1+x^{-1}u}{h}$, $\bar{g}=\dfrac{u(x^{-1}+u)}{6}$. Therefore,

$$t = \frac{u(x-x^{-1})}{h-\dfrac{1+x^{-1}u}{h}-\dfrac{x^{-1}+u}{x+u}\cdot\left(h-\dfrac{1+xu}{h}\right)} = \frac{u(x-x^{-1})}{\dfrac{(x-x^{-1})(h^2-1+u^2)}{h(x+u)}}$$

$$= \frac{uh(x+u)}{h^2-1+u^2} = \frac{uh(x+u)}{h^2+h} = \frac{u(x+u)}{h+1} = \frac{u(x+u)}{u^2} = \frac{x}{u}+1.$$

Since $\dfrac{t-a}{x} = \dfrac{t-1}{x} = \dfrac{1}{u} \in \mathbb{R}$, then $AT//OX$. Therefore, $AT \perp CD$. $\qquad\square$

2 Given integer $n \geq 3$. Does there exist an infinite number of sets each consisting of $2n$ positive integers that satisfy the following conditions?

(1) The greatest common divisor of all the elements in such a kind of set (denoted by A) is equal to 1;

(2) all the elements in A can be divided into two groups; each group contains n numbers that forms an arithmetic sequence, and the products of the numbers in each group are equal.

(Contributed by Yu Hongbin)

Solution The conclusion is that there certainly exists.

Let $a_k = r(a+(k-1)d)$, $b_k = s(a+kd)$, $k=1,2,\ldots,n$, where a,d,r,s are positive integers to be determined. It is easy to see that $a_1, a_2, \ldots, a_n$ and $b_1, b_2, \ldots, b_n$ form strictly increasing arithmetic sequences.

$a_1 a_2 \cdots a_n = b_1 b_2 \cdots b_n$ is equivalent to $r^n a = s^n(a+nd)$, namely,

$$(r^n - s^n)a = s^n dn. \qquad\qquad ①$$

Next we choose positive integers a, d, r, s in such a way that $a_1, a_2, \ldots, a_n, b_1, b_2, \ldots, b_n$ are not equal to each other and the greatest common divisor is 1.

For positive integer t, take $a = s = 1, r = tn + 1, d = \dfrac{r^n - 1}{n} = \dfrac{(tn + 1)^n - 1}{n}$. It is easy to find $r > n$, d is a positive integer and $\textcircled{1}$ holds.

Since $(b_1, b_2) = (1 + d, 1 + 2d) = (1 + d, d) = 1$, then the greatest common divisor of $a_1, a_2, \ldots, a_n, b_1, b_2, \ldots, b_n$ is 1. Furthermore, if there are two equal numbers in $a_1, a_2, \ldots, a_n, b_1, b_2, \ldots, b_n$, then it can only be $a_i = b_j$, where $i, j \in \{1, 2, \ldots, n\}$. Then

$$r(n + (i - 1)(r^n - 1)) = na_i = nb_j = n + j(r^n - 1).$$

Taking modulus of r on both sides shows that $0 \equiv n - j \pmod{r}$. Since $r = tn + 1 > n$, there is $j = n$. Substituting into the above equation and simplifying we get

$$n + (i - 1)(r^n - 1) = nr^{n-1}.$$

Taking modulus of r on both sides gives $n - i + 1 \equiv 0 \pmod{r}$. However, $0 < n - i + 1 \leq n < r, r \mid n - i + 1$ cannot be valid. Therefore, $a_1, a_2, \ldots, a_n, b_1, b_2, \ldots, b_n$ are not equal to each other. Set $A = a_1, a_2, \ldots, a_n, b_1, b_2, \ldots, b_n$ satisfies the given conditions. From the arbitrariness of t, we can construct an infinite number of sets A satisfying the requirements. $\square$

③ (1) Determine all positive integers n such that there exist n points $P_1, P_2, \ldots, P_n$ on the circumference of the unit circle, satisfying $|MP_1|^{2018} + |MP_2|^{2018} + \cdots + |MP_n|^{2018}$ is a constant when moving point M moves on the circumference.

    (2) Determine all positive integers n such that there exist n points $P_1, P_2, \ldots, P_n$ on the circumference of the unit circle, satisfying $|MP_1|^{2019} + |MP_2|^{2019} + \cdots + |MP_n|^{2019}$ is a constant when moving point M moves on the circumference. (Contributed by Wang Bin)

Solution Let $m = 1009$. We set the unit circumference in the problem to be the unit circumference in the complex plane, namely,

$$D = \{z \in \mathbb{C} : |z| = 1\}.$$

(1) Suppose $M, P_1, P_2, \ldots, P_n$ correspond to the complex numbers $z, w_1, w_2, \ldots, w_n \in D$, respectively. For $1 \leq l \leq n$, note that $z \cdot \bar{z} = w_l \cdot \bar{w}_l =$

1. Then

$$|MP_l|^2 = |z - w_l|^2 = \left|\frac{z}{w_l} - 1\right|^2 = \left(\frac{z}{w_l} - 1\right)\left(\frac{w_l}{z} - 1\right) = 2 - \frac{z}{w_l} - \frac{w_l}{z}.$$

Therefore,

$$|MP_l|^{2m} = \left(2 - \frac{z}{w_l} - \frac{w_l}{z}\right) m.$$

We consider function

$$g(x) = \left(2 - x - \frac{1}{x}\right)^m = \frac{(-1)^m (x-1)^{2m}}{x^m} = \sum_{k=-m}^{m} a_k x^k,$$

where the coefficient $a_k = (-1)^k C_{2m}^{m+k} \neq 0, k = -m, -m+1, \ldots, m$.

Let $S_k = w_1^k + w_2^k + \cdots + w_n^k, k \in \mathbb{Z}$. Assume that function $G(z) = \sum_{l=1}^{n} |MP_l|^{2m}$ takes a constant C on $M = z \in D$. Since

$$G(z) = \sum_{l=1}^{n} g\left(\frac{z}{w_l}\right) = \sum_{l=1}^{n} \sum_{k=-m}^{m} a_k z^k w_l^{-k}$$

$$= \sum_{k=-m}^{m} a_k \left(\sum_{l=1}^{n} w_l^{-k}\right) z^k = \sum_{k=-m}^{m} a_k S_{-k} z^k,$$

then the polynomial

$$(G(z) - C)z^m = \left(\left(\sum_{k=-m}^{m} a_k S_{-k} z^k\right) - C\right) z^m$$

of z is always zero on D, and thus is a zero polynomial.

Therefore, $a_k S_{-k} = 0$, $k = \pm 1, \pm 2, \ldots, \pm m$, and thus $S_1 = S_2 = \cdots = S_m = 0$.

If $n \leq m$, by Newton's identities, it is easy to find that all the primitive symmetric polynomials of $w_1, w_2, \ldots, w_n$ are zero, so $w_1 = w_2 = \cdots = w_n = 0$, a contradiction. Thus $n \geq m + 1$.

When $n \geq m + 1$, we can take $w_l = e^{\frac{2\pi i}{n}}, 1 \leq l \leq n$. They satisfy $S_{-k} = 0$, $k = \pm 1, \pm 2, \ldots, \pm m$, so $G(z)$ always takes a constant $C = n a_0 = n C_{2m}^m$ on D.

In conclusion, for (1), the required positive integer $n \geq 1010$.

(2) Without loss of generality, rotate properly so that $P_1, P_2, \ldots, P_n$ is all in set $A = \{e^{i\theta} \mid 2\varepsilon < \theta < 2\pi\}$. Here we arbitrarily take $0 < \varepsilon < \frac{\pi}{n}$.

For $1 \le l \le n$, we can let $P_l = w_l^2$, where $w_l \in A_1 = \{e^{i\theta} \mid \varepsilon < \theta < \pi\}$. We take

$$B = D \backslash A = \{z \in D \mid 0 \le \arg z \le 2\varepsilon\}, \quad B_1 = \{z \in D \mid 0 \le \arg z \le \varepsilon\}.$$

For any $z \in B_1$, there is $z^2 \in B$. Suppose the corresponding point of z^2 is M, then for $1 \le l \le n$ there is

$$|MP_l|^2 = \left|z^2 - w_l^2\right|^2 = 2 - \frac{z^2}{w_l^2} - \frac{w_l^2}{z^2} = \left(\left(\frac{z}{w_l} - \frac{w_l}{z}\right)i\right)^2.$$

Since $0 \le \arg z \le \varepsilon < \arg w_l < \pi$, then $\dfrac{z}{w_l}$ is at the lower semicircle of D, and $\dfrac{w_l}{z}, \dfrac{z}{w_l}$ are conjugate to each other, so $\dfrac{z}{w_l} - \dfrac{w_l}{z}$ is a pure imaginary number with negative imaginary part, i.e. $\left(\dfrac{z}{w_l} - \dfrac{w_l}{z}\right)i$ is a positive real number. Therefore,

$$|MP_l| = \left(\frac{z}{w_l} - \frac{w_l}{z}\right)i.$$

Then

$$|MP_l|^{2m+1} = \left(\frac{z}{w_l} - \frac{w_l}{z}\right)^{2m+1} i^{2m+1}.$$

Consider function

$$h(x) = i^{2m+1}\left(x - \frac{1}{x}\right)^{2m+1} = i^{2m+1}\frac{(x^2 - 1)^{2m+1}}{x^{2m+1}}$$

$$= \sum_{k=0}^{2m+1} b_k x^{2k-2m-1},$$

where the coefficient $b_k = i^{2m+1}(-1)^{k+1}C_{2m+1}^k \neq 0, 0 \le k \le 2m + 1$.

Assume that function $H(z) = \sum_{l=1}^{n} |MP_l|^{2m+1}$ is always a constant C when $z \in B_1$. Since

$$H(z) = \sum_{l=1}^{n} h\left(\frac{z}{w_l}\right) = \sum_{l=1}^{n} \sum_{k=0}^{2m+1} b_k Z^{2k-2n-1} w_l^{2n+1-2k}$$

$$= \sum_{k=0}^{2m+1} b_k \left(\sum_{l=1}^{n} w_l^{2m+1-2k}\right) Z^{2k-2m-1}$$

$$= \sum_{k=0}^{2m+1} b_k S_{2m+1-2k} Z^{2k-2m-1},$$

then the polynomial $(H(Z) - C)Z^{2m+1} = (\sum_{k=0}^{2m+1} b_k S_{2m+1-2k} Z^{2k}) - CZ^{2m+1}$ of z is always zero on B_1, and thus is a zero polynomial. In particular, for the coefficient z^{2m} we have $b_m S_1 = 0$, i.e. $S_1 = w_1 + w_2 + \cdots + w_n = 0$. However, $\mathrm{Im} S_1 = \sum_{l=1}^{n} \mathrm{Im}(w_l) > 0$, and this is a contradiction.

Consequently, for question (2) there exists no positive integer n that satisfies the given conditions. $\qquad \square$

Second Day
(8:00 – 12:30; March 5, 2019)

4 We call an infinite sequence of positive integers $a_1, a_2, \cdots$ a "good sequence", if for any different positive integers m, n, we have

$$(m, n) \left| a_m^2 + a_n^2, (a_m, a_n) \right| m^2 + n^2,$$

where (a, b) denotes the greatest common divisor of positive integers a, b. For positive integers a, k, we call a is " k-good" if there exists a good sequence $a_1, a_2, \cdots$ satisfying $a_k = a$.

Do there exist positive integers k such that there are exactly 2019 positive integers that are k-good? Prove your conclusion. (Contributed by He Yijie)

Solution   Denote the whole of the good sequences as Γ.
We first prove the following lemma.

Lemma   *A necessary and sufficient condition for a positive integer sequence $\{a_n\} \in \Gamma$ is that there is $m \mid a_m^2, a_m \mid m^2$ for any positive integer m.*

Proof of the lemma   First, we prove the necessity. Suppose sequence of positive integers $\{a_n\} \in \Gamma$. By the condition, we know that for any positive integer m, there is

$$(m, 2m) \left| a_m^2 + a_{2m}^2, (2m, 3m) \right| a_{2m}^2 + a_{3m}^2, (3m, m) \mid a_{3m}^2 + a_m^2.$$

Since $(m, 2m) = (2m, 3m) = (3m, m) = m$, then

$$m \mid (a_m^2 + a_{2m}^2) + (a_{2m}^2 + a_{3m}^2) - (a_{3m}^2 + a_m^2),$$

namely, $m \mid 2a_m^2$. Substituting $2m$ for m, we get $2m \mid 2a_{2m}^2$, namely, $m \mid a_{2m}^2$. Then, from $m \mid a_m^2 + a_{2m}^2$, we obtain $m \mid a_m^2$.

In order to prove $a_m \mid m^2$, it is only necessary to show that for any prime p, there is $v_p(a_m) \leq 2v_p(m)$. In fact, for any positive integer $m \neq n$, from $(a_m, a_n) \mid m^2 + n^2$ we obtain

$$\min\{v_p(a_m), v_p(a_n)\} \leq v_p(m^2 + n^2).$$

Denote $\alpha = v_p(m)$. Let $n = p^{4\alpha+1}$, so $n \neq m$. $v_p(m^2) = 2\alpha < v_p(n^2)$, and thus

$$v_p(m^2 + n^2) = 2\alpha.$$

Therefore,

$$\min\{v_p(a_m), v_p(a_n)\} \leq 2\alpha.$$

According to the proof above, we know that $n \mid a_n^2$, so $2v_p(a_n) \geq 4\alpha + 1$, and hence $v_p(a_n) \geq 2\alpha + 1$. Then we find

$$v_p(a_m) \leq 2\alpha = 2v_p(m),$$

and thus $a_m \mid m^2$. We finish the proof of the necessity.

Next, we prove the sufficiency. Suppose positive integer sequence $\{a_n\}$ satisfies: for any positive integer m, there is $m \mid a_m^2, a_m \mid m^2$. Then for any positive integers $m \neq n$, from $m \mid a_m^2, n \mid a_n^2, a_m \mid m^2, a_n \mid n^2$, we have $(m, n) \mid a_m^2 + a_n^2$, $(a_m, a_n) \mid m^2 + n^2$. Therefore, $\{a_n\} \in \Gamma$. The proof of the sufficiency is complete.

We return to the original question. Let the standard decomposition be $k = \prod_{i=1}^{t} p_i^{\alpha_i}$. From the lemma we learn that a_k has the same set of prime factors as that of k when $\{a_n\} \in \Gamma$. Let $a_k = \prod_{i=1}^{t} p_i^{\beta_i}$. Then for each $i(1 \leq i \leq t)$, the necessary and sufficient conditions for α_i, β_i to be satisfied are $\beta \leq 2\alpha_i, \alpha_i \leq 2\beta_i$. Therefore, after α_i is given, the number of ways to take β_i is $2\alpha_i - \left[\dfrac{\alpha_i - 1}{2}\right] = \dfrac{3\alpha_i + \varepsilon_i}{2}$, where $\varepsilon_i = 1$ when α_i is odd and $\varepsilon_i = 2$ when α_i is even.

Therefore, the number of ways to take array $(\beta_1, \beta_2, \ldots, \beta_t)$ is $\prod_{i=1}^{t} \dfrac{3\alpha_i + \varepsilon_i}{2}$. Notice that positive integers a_k corresponding to different $(\beta_1, \beta_2, \ldots, \beta_t)$ are different from each other, and that these a_k are indeed k-good. For example, the kth term of a good sequence takes that a_k value and the other terms are $a_l = l, l \neq k$. Therefore, the number N of k-good is $N_k = \prod_{i=1}^{t} \dfrac{3\alpha_i + \varepsilon_i}{2}$. It is obvious that N_k is not a multiple of 3, and thus $N_k \neq 2019$.

Consequently, there does not exist positive integer k that satisfies the condition. $\square$

5 Let $\mathbb{Q}$ be the set of all rational numbers. Find all functions $f : \mathbb{Q} \to \mathbb{Q}$ that satisfy, for any rational numbers x, y,

$$4f(x)f(y) + \frac{1}{2} = f\left(2xy + \frac{1}{2}\right) + f(x - y).$$

(Contributed by Lin Bo)

Solution In the original equation let $y = 0$, and then we get

$$(4f(0) - 1)f(x) = f\left(\frac{1}{2}\right) - \frac{1}{2}.$$

It is clear that f is not a constant function (Otherwise, assume $f(x) = c$, then $4c^2 + \frac{1}{2} = 2c$ has no real number solution). From the above formula we have $4f(0) - 1 = 0, f\left(\frac{1}{2}\right) = \frac{1}{2}$.

In the original equation let $y = \frac{1}{2}$, then

$$2f(x) + \frac{1}{2} = f\left(x + \frac{1}{2}\right) + f\left(x - \frac{1}{2}\right).$$

Let $g(x) = f\left(x + \frac{1}{2}\right) - f(x)$. Then for any rational number x, there is

$$g\left(x + \frac{1}{2}\right) = g(x) + \frac{1}{2}.$$

By exchanging x, y in the original equation and comparing it with the original equation, we immediately know that f is an even function.

Substitute $-y$ for y in the original equation, and compare with the original equation (note that f is an even function), we get

$$f\left(2xy + \frac{1}{2}\right) + f(x - y) = f\left(-2xy + \frac{1}{2}\right) + f(x + y).$$

Hence

$$f(x + y) - f(x - y) = f\left(2xy + \frac{1}{2}\right) - f\left(-2xy + \frac{1}{2}\right)$$

$$= \left(f\left(2xy + \frac{1}{2}\right) - f(2xy)\right)$$

$$+ \left(f(2xy) - f\left(2xy - \frac{1}{2}\right)\right)$$

$$= g(2xy) + g\left(2xy - \frac{1}{2}\right) = 2g(2xy) - \frac{1}{2}.$$

Let $h(x) = g(x) - \dfrac{1}{4}$. Then the above equation becomes

$$f(x+y) - f(x-y) = 2h(2xy). \qquad ①$$

For any positive integer n, let $y = \dfrac{n}{2}$ in $①$, and then

$$2h(nx) = f\left(x + \frac{n}{2}\right) - f\left(x - \frac{n}{2}\right) = \sum_{i=-n}^{n-1} \left(f\left(x + \frac{i+1}{2}\right) - f\left(x + \frac{i}{2}\right)\right)$$

$$= \sum_{i=-n}^{n-1} g\left(x + \frac{i}{2}\right) = \sum_{i=-n}^{n-1} \left(g(x) + \frac{i}{2}\right) = 2ng(x) - \frac{n}{2} = 2nh(x).$$

Substituting $-y$ for y in $①$, we know that h is an odd function. Then for any integer $-y$ and rational number x, there is $h(nx) = nh(x)$. Thus for any rational number r, x, there is $h(rx) = rh(x)$. In fact, we can let $r = \dfrac{q}{p}$, where p, q are integers and $p \neq 0$. Then

$$p \cdot h\left(\frac{q}{p}x\right) = h(qx) = qh(x).$$

Hence $h\left(\dfrac{q}{p}x\right) = \dfrac{q}{p} \cdot h(x)$. Next taking $x = 1$, we have $h(r) = rh(1) = 2rh\left(\dfrac{1}{2}\right)$.

Now we calculate $h\left(\dfrac{1}{2}\right)$. By the definition,

$$h\left(\frac{1}{2}\right) = g\left(\frac{1}{2}\right) - \frac{1}{4} = f(1) - f\left(\frac{1}{2}\right) - \frac{1}{4} = f(1) - \frac{3}{4}.$$

In the original equation, let $x = y = \dfrac{1}{2}$. Then $f(1) = \dfrac{5}{4}$, and thus $h\left(\dfrac{1}{2}\right) = \dfrac{1}{2}$. Therefore, $h(x) = x$.

Finally, let $y = x$ in $①$. Then we have $f(2x) = 2h(2x^2) + f(0) = 4x^2 + \dfrac{1}{4}$. Therefore,

$$f(x) = x^2 + \frac{1}{4}.$$

It is checked that the above function satisfies the original equation.

In summary, the original equation has a unique solution $f(x) = x^2 + \dfrac{1}{4}$. □

6 Alice and Bob play the following game: Set a positive real number k, and 80 numbers on the circumference of a circle in succession, which are all equal to 0 at the beginning. The game is played one round at a time, and each round is played with the following two steps in turn:

Step 1: Alice adds a non-negative real number to each number on the circumference so that the sum of the 80 numbers increases by 1.

Step 2: Bob selects 10 adjacent numbers on the circumference that have the largest sum (if there are more than one set of 10 adjacent numbers reaching the largest value, then choose any one among them.), and changes all of them into 0.

At the end of each round, if there is a number on the circumference that is not less than k, then Alice wins, and the game is over. Otherwise, the next round will go on.

Find all the positive real numbers k such that Alice has a winning strategy. (Contributed by Qu Zhenhua)

Solution Alice has a winning strategy if an only if $k < 1 + \dfrac{1}{2} + \dfrac{1}{3} + \cdots + \dfrac{1}{7} = \dfrac{363}{140}$.

More generally, let integers $m, n > 1$, the same game be played over mn numbers, and each time Bob change the largest sum of m consecutive numbers to 0. Then Alice has a winning strategy if and only if $k < H(n - 1) = 1 + \dfrac{1}{2} + \cdots + \dfrac{1}{n-1}$.

Firstly, we will prove that if $k < H(n - 1)$, then Alice has a winning strategy.

Take any positive real number $\varepsilon < 0.1$ and $\varepsilon < H(n - 1) - k$. Let s be the smallest positive integer such that $s \cdot \dfrac{\varepsilon}{2(n - 1)} \geq 1 - \varepsilon$. The mn numbers are denoted in order as $x_1, x_2, \ldots, x_{mn}$. In the first s rounds, Alice increases on x_m by $1 - \dfrac{\varepsilon}{2}$, and on each of $x_{2m}, x_{3m}, \ldots, x_{mm}$ by $\dfrac{\varepsilon}{2(n - 1)}$. At this time, because the sum of any m consecutive numbers that contain x_m is the largest, so each time Bob will change x_m into 0. After s rounds, there is

$$2 \leq i \leq n, x_{im} = s \cdot \dfrac{\varepsilon}{2(n - 1)},$$

while the rest of numbers are all 0. Note that by the definition of s we have

$$1 - \varepsilon \leq s \cdot \dfrac{\varepsilon}{2(n - 1)} < 1 - \dfrac{2n - 3}{2n - 2}\varepsilon.$$

In round $s+1$, Alice adds $\dfrac{1}{n-1}$ to each of the $n-1$ numbers greater than 0, while Bob will change one of them into 0. In round $s+2$, Alice adds $\dfrac{1}{n-2}$ to each of the remaining $n-2$ numbers greater than 0, and Bob will change one of them into 0, and so on. In round $s+(n-2)$, Alice adds $\dfrac{1}{2}$ to each of the 2 numbers greater than 0, and Bob changes one of them into 0. At the end of this round, the remaining number that is greater than 0 is

$$s \cdot \frac{\varepsilon}{2(n-1)} + \frac{1}{n-1} + \frac{1}{n-2} + \cdots + \frac{1}{2} \geq H(n-1) - \varepsilon > k.$$

Hence, Alice wins.

Next, we will prove that Alice cannot win when $k \geq H(n-1)$. That is, no matter what Alice does, there will be no number $\geq H(n-1)$ at the end of each round. We use the proof by contradiction. Assume that there is a number not less than $H(n-1)$ at the end of the Nth round. We call a group of i consecutive numbers as a small block, where $1 \leq i \leq m$. Two small blocks are said to be non-intersecting if they do not share a number in the same position. A small block is said to be non-zero if both numbers at the two ends of the block are non-zero. Then at the end of the $N\text{th}$ round, there is a non-zero small block in which the sum of the numbers is not less than $H(n-1)$.

For $t = 0, 1, 2, \ldots, n-2$, we will prove by induction that at the end of the $(N-t)$th round, there are no more than $t+1$ non-intersecting non-zero small blocks, with the sum of all the numbers in these blocks not less than $(t+1)(H(n-1) - \sum_{i=2}^{t+1} \frac{1}{i})$ (When $t = 0$, the last summation is considered to be 0).

When $t = 0$, i.e., at the end of the Nth round, the conclusion certainly holds.

Assume that the conclusion holds at the end of the $(N-t)$th round, for $t < n-2$. Then non-intersecting non-zero small blocks $A_1, A_2, \ldots, A_q, q \leqslant t+1$ can be found, such that

$$\sum_{i=1}^{q} \sigma(A_i) \geq (t+1)\left(H(n-1) - \sum_{i=2}^{t+1} \frac{1}{i}\right),$$

where $\sigma(A_i)$ denotes the sum of the numbers in A_i. Since there exist numbers greater than zero, then $N-t > 0$. In Step 2 of round $N-t$, Bob changes m consecutive numbers $x_s, x_{s+1}, \ldots, x_{s+m-1}$ into 0. Since he must choose those m adjacent numbers with the largest sum, so before the change made by Bob, the sum of $x_s, x_{s+1}, \ldots, x_{s+m-1}$ is not less than

$$\frac{1}{q} \sum_{i=1}^{q} \sigma(A_i) \geq H(n-1) - \sum_{i=2}^{t+1} \frac{1}{i}.$$

At the end of Step 1 in the $(N-t)$th round, let A_{q+1} be the nonzero small block containing all nonzero numbers in $x_s, x_{s+1}, \ldots, x_{s+m-1}$, then A_{q+1} does not intersect with $A_1, \ldots, A_q$, and

$$\sum_{i=1}^{q+1} \sigma(A_i) \geq (t+2) \left(H(n-1) - \sum_{i=2}^{t+1} \frac{1}{i} \right).$$

Subtracting what Alice adds to each number in the circumference in Step 1, then the sum of all the numbers of $A_1, A_2, \ldots, A_{q+1}$ is at most reduced by 1. At this point, these small blocks are then appropriately shortened into smaller non-zero blocks (it is even possible that the entire small block be deleted), and we get non-intersecting non-zero small blocks $B_1, B_2, \ldots, B_p, p \leq q+1 \leq t+2$, and

$$\sum_{i=1}^{p} \sigma(B_i) \geq \left(\sum_{i=1}^{q+1} A_i \right) - 1$$

$$\geq (t+2) \left(H(n-1) - \sum_{i=2}^{t+1} \frac{1}{i} \right) - 1$$

$$= (t+2) \left(H(n-1) - \sum_{i=2}^{t+2} \frac{1}{i} \right).$$

Therefore, at the end of round $N-t-1$, the conclusion also holds. By induction, we know that at the end of the $N-(n-2)$th round, there exist no more than $n-1$ non-intersecting non-zero small blocks in which the sum of the numbers is not less than

$$(n-1) \left(H(n-1) - \sum_{i=2}^{n-1} \frac{1}{i} \right) = n-1.$$

Let the sum of all numbers at the end of round i be S_i, then $S_0 = 0$. If $S_i < n-1$, then in round $i+1$ the sum of all numbers after the operation of Alice is $S_i + 1$. However, Bob changes the number on m adjacent vertices with the largest sum to 0, and the sum of these numbers is at least $\dfrac{1}{n}(S_i+1)$. Therefore,

$$S_{i+1} \leq \frac{n-1}{n}(S_i + 1) < n - 1.$$

Thus by mathematical induction, we have $S_i < n - 1$ for all i, which contradicts $S_{N-(n-2)} \geq n - 1$. Therefore, Alice cannot win when $k \geq H(n - 1)$.

The original problem corresponds to the case $m = 10$ and $n = 8$, so that Alice has a winning strategy if and only if $k < H(7) = \dfrac{363}{140}$. $\square$

Test II, First Day
(8:00 – 12:30; March 9, 2019)

1 Suppose point A lies outside circle ω, lines AB, AC are tangent to ω at points B, C, respectively, and point P is a moving point on the minor arc $\overset{\frown}{BC}$, not including B, C. Construct a tangent line of ω through P, intersecting with AB, AC at points D, E, respectively. Lines BP, CP intersect the interior angle bisector of $\angle BAC$ at points U, V, respectively. Make vertical line of AB through P, intersecting with line DV at point M. Make vertical line of AC through P, intersecting with line EU at point N.

Prove that there exists a fixed point L independent of point P such that points M, N, L are collinear. (Contributed by Zhang Sihui)

Solution   Let us first prove the following lemma.

Lemma   *In quadrilateral $XYZW, XW//YZ$, points S, T are on XY, ZW, respectively. If $XT//SZ$, then $WS//TY$.*

Proof of the lemma   If $XY//ZW$, then quadrilaterals $XYZW, XSZT$ are both parallelograms, and thus

$$YS = XY - XS = ZW - ZT = WT.$$

Therefore, quadrilateral $WSYT$ is also a parallelogram, and thus $WS//TY$.

If XY, ZW are not parallel, then let the lines in which they are located intersect at point K. From $XT//YZ, XW//SZ$, we have

$$\frac{KX}{KY} = \frac{KW}{KZ}, \frac{KX}{KS} = \frac{KT}{KZ}.$$

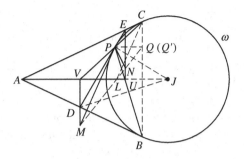

Fig. 1.1

Dividing the two equations, gives $\dfrac{KS}{KY} = \dfrac{KW}{KT}$, and thus $WS//TY$. The lemma is proved.

Return to the original question. As shown in the Fig. 1.1, let L be the orthocenter of $\triangle ABC$. In the following, we will prove that points L, M, N are collinear. It is only necessary to consider the case that L, M, N do not coincide.

Let the center of ω be J. Then J is the A-excentre of $\triangle ADE$. Since

$$\angle PVJ = \angle CAV + \angle ACV$$

$$= \frac{1}{2}\angle BAC + \frac{1}{2}\angle AED$$

$$= \frac{1}{2}\angle EDB = \angle PDJ,$$

then P, J, D, V are concyclic, and hence $\angle JVD = \angle JPD = 90°$, namely, $DV\perp AJ$. Suppose lines ML, BC intersect at Q. Since VM, CQ are both perpendicular to AJ, then $VM//CQ$. And as CL, PM are both perpendicular to AB, then $CL//PM$. Applying the lemma to the quadrilateral $CVMQ$ and points P, L, we know that $PQ//VL$, i.e., $PQ\perp BC$.

Suppose lines NL, BC intersect at point Q'. Similarly, we obtain $PQ'\perp PC$. Therefore, $Q = Q'$. Consequently, points M, N, L, Q are concyclic. In particular, M, N, L are collinear. The conclusion is confirmed. $\square$

2 Let S be the set of all the arrays of 10 non-negative integers $(x_1, x_2, \ldots, x_{10})$ satisfying $x_1 + x_2 + \cdots + x_{10} = 2019$. An operation on an array in S is defined as: Select a component that is not less than 9, subtract 9 from this component, and add 1 to each of the other components. For arrays $A, B \in S$, $A \to B$ indicates that array A can be turned into array B after a number of the operations. For integer $k \geq 0$, denote

$$S_k = \{(x_1, x_2, \ldots, x_{10}) \in S \mid \min_{1 \leq i \leq 10} x_i \geq k\}.$$

(1) Find the minimum k with the following property: for any $A, B \in S_k$, if $A \to B$, then $B \to A$.

(2) Let k be the value determined in (1). What is the maximum number of arrays that can be taken out from S_k, such that in these arrays, any one cannot be turned into another one by operations? (Contributed by Qu Zhenhua)

Solution 1   We denote the operation of subtracting 9 from the ith component and adding 1 to each of the remaining 9 components as T_i. We consider the following two arrays:

$$A = (7, 7, \ldots, 7, 1956), B = (8, 8, \ldots, 8, 1947).$$

Doing T_{10} on A gives you B, but it is impossible to do a series of operations on B to get A. In fact, if it is possible, then there exist sequence $B_0 = B, B_1, \ldots, B_n = A$, where for each $1 \leq i \leq n$, B_i is obtained by the operation on B_{i-1}. Since there are 9 components with number 7 in B_n, then there are at least 8 components with number 6 in B_{n-1}, and there are at least 7 component with number 5 in B_{n-2}. Considering by successive steps, we can see that there are at least 2 commpontents with number 0 in B_{n-7}. But it is impposible to do an operation on B_{n-8} to get an array that has two components with number 0. so $k \geq 8$.

In the following we will prove that $k = 8$ satisfies the requirement. For $A = (x_1, x_2, \ldots, x_{10}) \in S$, define

$$f(A) = (x_1 - x_2, x_1 - x_3, \ldots, x_1 - x_{10})(\bmod 10).$$

If operating on array A once gives array B, it is easy to know that $f(A) = f(B)$. Thus for $A, B \in S_8$, if $f(A) \neq f(B)$, then $A \nrightarrow B$.

In the following we will prove that if $A, B \in S_8$ and $f(A) = f(B)$, then $A \to B$.

First note that, for any $\alpha = (x_1, x_2, \cdots, x_{10}) \in S_8$, and $i, j \in \{1, 2, \cdots, 10\}, i \neq j$, if $x_i \geq 18$, then after several operations x_i can be decreased by 10, x_j increased by 10, and the other components remain unchanged. In fact, it is enough to do operations $T_i, T_{i_1}, T_{i_2}, \ldots, T_{i_8}, T_i$ in order, where $i_1, \ldots, i_8$ are the remaining 8 positions except i, j. We will denote such a series of operations as $T_{i,j}$.

Now we consider $A, B \in S_8$, $f(A) = f(B)$. Let $A = (x_1, x_2, \ldots, x_{10})$ and $B = (y_1, y_2, \ldots, y_{10})$. Take any subscript i satisfying $x_i \geq 200$. After doing no more than 9 times operation T_i, A becomes $A_1 = (z_1, z_2, \ldots, z_{10})$ and $z_i \equiv y_i \pmod{10}$. Then by $f(A_1) = f(A) = f(B)$ we know that $A_1 \equiv B \pmod{10}$.

If $A_1 \neq B$, then there exist subscripts $i, j, i \neq j$ such that $z_i > y_i, z_j < y_j$. Do operation $T_{i,j}$, which keeps each component mod 10 a constant. Continue this process. Since $\sum_{i=1}^{10} |z_i - y_i|$ is strictly decreasing, then it is possible to change A_1 to B after several operations.

Finally, since the sum of the 9 components in $f(A)$ is

$$9x_1 - x_2 - x_3 - \cdots - x_{10} = 10x_1 - (x_1 + x_2 + \cdots + x_{10})$$

$$\equiv -2019 \pmod{10},$$

then in $f(A)$ after its first 8 components mod 10 is determined, the 9th component mod 10 is also uniquely determined. And in $f(A)$ there are 10^8 possibilities for choosing its first 8 components mod10. For example, for the given $a_1, \ldots, a_8 \in \{0, 1, \ldots, 9\}$, taking

$$A = (10, 20 - a_1, 20 - a_2, \ldots, 20 - a_8, b) \in S_8,$$

then $f(A) = (a_1, a_2, \ldots, a_8, 10 - b) \pmod{10}$. Therefore, there are at most 10^8 arrays that can be taken out in S_8, in which any one cannot be obtained by doing operations on another one. Furthermore, we also proved that for any $A, B \in S_8$, if $A \to B$, then $B \to A$. $\qquad\square$

Solution 2 An alternative proof of the following conclusion is given here, and the rest is the same as in Solution 1.

We will prove that if A, B are both arrays in S_8, satisfying

$$x_1 - y_1 \equiv x_2 - y_2 \equiv \cdots \equiv x_{10} - y_{10} \pmod{10}, \qquad (1)$$

then $A \to B$.

We still denote the operation of subtracting 9 from the ith component and adding 1 to each of the remaining 9 components as T_i. First, we show that there exists non-negative integers $a_1, a_2, \ldots, a_{10}$ such that, after doing a_1 times T_1, a_2 times $T_2, \ldots, a_{10}$ times T_{10}, A becomes B. If we ignore the order of these operations and allow negative number components appear in an array during the operations, the proof will be equivalent to proving that the following equation system

$$x_i - 9a_i + \sum_{j \neq i} a_j = y_i, i = 1, 2, \ldots, 10 \qquad (2)$$

has non-negative integer solutions $a_1, a_2, \ldots, a_{10}$. Let $S = a_1 + a_2 + \cdots + a_{10}$. The solution of $\textcircled{2}$ is

$$a_i = \frac{1}{10}(x_i - y_i + S)i = 1, 2, \cdots, 10. \qquad \textcircled{3}$$

Take a sufficiently large S satisfying $1 \leq i \leq 10, x_i - y_i + S \geq 0$ and $10 \mid x_i - y_i + S$. Note that by $\textcircled{1}$, only $10 \mid x_1 - y_1 + S$ is required, then for all i, there is $x_i - y_i + S$. Thus, $a_1, a_2, \cdots, a_{10}$ defined in $\textcircled{3}$ are the non-negative integer solution of the equation system $\textcircled{2}$.

Then we need to show that these $a_1 + a_2 + \cdots + a_{10}$ operations can be arranged in proper order, so that no negative coordinates will appear during the operations. It is sufficient to state that if the coordinates of A are all non-negative after a number of operations, there must be one among the remaining operations that, when be executed, will keep the non-negativity of the coordinates.

We use proof by contradiction. Assume that, after a series of operations, array A becomes $(z_1, z_2, \ldots, z_{10})$, on which each of the remaining operations will produce negative coordinates. Since there must be some $z_i \geq 9$, the remaining operations cannot contains every $T_1, T_2, \cdots, T_{10}$. It may be assumed that they include $T_1, \ldots, T_m$, and not includ $T_{m+1}, \ldots, T_{10}$, where $1 \leq m \leq 9$. Therefore, $z_1, \ldots, z_m \leq 8$. Suppose T_i still need to be executed b_i times, $1 \leq i \leq m$, and $b_1 = \max(b_1, \cdots, b_m)$. Since after all these remaining operations done (without considering whether the intermediate processes produce negative coordinates) the result is B. Therefore,

$$y_1 = z_1 - 9b_1 + b_2 + \cdots + b_m \leq z_1 - b_1 < z_1 \leq 8.$$

This is in contradiction to $B \in S_8$. □

3 Let n be a given positive even number and $a_1, a_2, \ldots, a_n$ be the n non-negative real numbers whose sum is 1. Find the maximum possible value of

$$S = \sum_{1 \leq i < j \leq n} \min\{(j - i)^2, (n - j + i)^2\} a_i a_j.$$

(Contributed by Fu Yunhao)

Solution 1 Let $n = 2k$. When $a_1 = a_{k+1} = \dfrac{1}{2}$ and the remaining a_i are 0, $S = \dfrac{n^2}{16}$. In the following we will show that this is the maximum possible value of S, i.e., there is always $S \leq \dfrac{n^2}{16}$.

Suppose $\varepsilon = e^{\frac{2\pi i}{n}}$ is a unit root of order n. Let

$$z = a_1\varepsilon + a_2\varepsilon^2 + \cdots + a_n\varepsilon^n.$$

Then,

$$0 \le |z|^2 = z \cdot \bar{z} = (a_1\varepsilon + a_2\varepsilon^2 + \cdots + a_n\varepsilon^n)(a_1\varepsilon^{-1} + a_2\varepsilon^{-2} + \cdots + a_n\varepsilon^{-n})$$

$$= \sum_{i=1}^{n} a_i^2 + \sum_{1 \le i < j \le n} a_ia_j(\varepsilon^{i-j} + \varepsilon^{j-i})$$

$$= \sum_{i=1}^{n} a_i^2 + \sum_{1 \le i < j \le n} 2\cos\frac{2(j-i)\pi}{n}a_ia_j$$

$$= \left(\sum_{i=1}^{n} a_i\right)^2 - 4\sum_{1 \le i < j \le n} \sin^2\frac{(j-i)\pi}{n}a_ia_j$$

$$= 1 - 4\sum_{1 \le i < j \le n} \sin^2\frac{(j-i)\pi}{n}a_ia_j.$$

Therefore,

$$\sum_{1 \le i < j \le n} \sin^2\frac{(j-i)\pi}{n}a_ia_j \le \frac{1}{4}.$$

We only need to prove that for all $1 \le i < j \le n$, there is

$$\frac{n^2}{4}\sin^2\frac{(j-i)\pi}{n} \ge \min\{(j-i)^2, (n-j+i)^2\},$$

or equivalently,

$$\sin\frac{(j-i)\pi}{n} \ge \frac{2}{n}\min\{j-i, n-j+i\}.$$

Since $\sin\dfrac{(j-i)\pi}{n} = \sin\dfrac{(n-j+i)\pi}{n}$, we only need to consider the case $0 < j - i \le \dfrac{n}{2}$. Let $\dfrac{j-i}{n} = x \le \dfrac{1}{2}$. We will show that $\sin(x\pi) \ge 2x$ holds for $x \in \left[0, \dfrac{1}{2}\right]$.

Let $f(x) = \sin(x\pi) - 2x$. Since $f'(x) = \pi\cos(x\pi) - 2$ decreases on $\left[0, \dfrac{1}{2}\right]$ and has a unique zero point, then $f(x)$ increases and then decreases on $\left[0, \dfrac{1}{2}\right]$. Combining $f(0) = f\left(\dfrac{1}{2}\right) = 0$ shows that $f(x) \ge 0$ holds only for $x \in \left[0, \dfrac{1}{2}\right]$.

Solution 2 Here we give another proof of $S \leq \dfrac{n^2}{16}$. Let $n = 2k$. For $1 \leq m \leq k$, using the inequality of arithmetic and geometric means, we have

$$\sum_{i=m}^{m+k-1} a_i \cdot \sum_{j=m+k}^{m+2k-1} a_j \leq \left(\frac{a_1 + a_2 + \cdots + a_n}{2}\right)^2 = \frac{1}{4},$$

where the subscript is understood as mod n. Summing the above inequality over m,

$$\sum_{m=1}^{k} \left(\sum_{i=m}^{m+k-1} a_i \cdot \sum_{j=m+k}^{m+2k-1} a_j\right) \leq \frac{k}{4}.$$

Expanding the left side, for $1 \leq i < j \leq n$, $a_i a_j$ appears exactly $\min\{j - i, n - j + i\}$ times. Therefore,

$$\sum_{m=1}^{k} \left(\sum_{i=m}^{m+k-1} a_i \cdot \sum_{j=m+k}^{m+2k-1} a_j\right) = \sum_{1 \leq i < j \leq n} \min\{j - i, n - j + i\} a_i a_j \leq \frac{k}{4}.$$

And since $\min\{j - i, n - j + i\} \leq k$, then

$$\min\{(j - i)^2, (n - j + i)^2\}$$
$$= (\min\{j - i, n - j + i\})^2 \leq k \min\{j - i, n - j + i\}.$$

Consequently,

$$S = \sum_{1 \leq i < j \leq n} \min\{(j - i)^2, \ (n - j + i)^2\} a_i a_j$$

$$\leq \sum_{1 \leq i < j \leq n} k \min\{j - i, n - j + i\} a_i a_j$$

$$\leq \frac{k^2}{4} = \frac{n^2}{16}.$$

Solution 3 (By Yu Ranfeng) A method of adjustment to prove $S \leq \dfrac{n^2}{16}$ is given here. We still let $n = 2k$.

Let $a_1, a_2, \ldots, a_n$ be uniformly distributed on a circumference with length n. Then $\min\{j - i, n - j + i\}$ is exactly the length of the minor arc between a_i and a_j. It follows that the expression of S has permutation invariance.

If there are three non-zero numbers on a semicircle, we may assume they are $a_1, a_i, a_j > 0, 1 < i < j \leq k + 1$. We substitute

$$a'_1 = a_1 - (j - i)x, a'_i = a_i + (j - 1)x, a'_j = a_j - (i - 1)x$$

for a_1, a_i, a_j, respectively, where x is to be determined. Note that

$$a'_1 + a'_i + a'_j = a_1 + a_i + a_j.$$

Now S is considered as a function of x, denoted by $S(x)$, which is a polynomial function of x at most degree 2. Since the x^2 term appears only in the pairwise product of a'_1, a'_i, a'_j, with the coefficient

$$-(j - i)(j - 1)(i - 1)^2 + (j - i)(i - 1)(j - 1)^2$$
$$-(j - 1)(i - 1)(j - i)^2 = 0,$$

$S(x)$ is then a linear polynomial of x.

The range of x in which a'_1, a'_i, a'_j keep non-negative is the closed interval $\left[-\dfrac{a_i}{j - 1}, \min\left(\dfrac{a_1}{j - i}, \dfrac{a_j}{i - 1} \right) \right]$. Therefore $S(x)$ reaches its maximum value at some endpoint and its value is not less than the original $S = S(0)$. After such an adjustment, one of a'_1, a'_i, a'_j becomes zero, i.e., the number of zeros in $a_1, a_2, \ldots, a_n$ is increased by 1.

After this adjustment, if there are still more than four non-zero numbers, three of them must be on the same semicircle, and the adjustment can be continued.

If only one number on the circumference is non-zero, then $S = 0$.

If there are only two numbers on the circumference that are non-zero, let them be $a_1, a_i, 1 < i \leq k + 1$, then

$$S = (i - 1)^2 a_1 a_i \leq k^2 a_1 a_i \leq \frac{k^2}{4} = \frac{n^2}{16}.$$

If there are only three non-zero numbers on the circumference, then the positions of these three numbers on the circumference of the circle must be the three vertices of an acute triangle. Suppose these three numbers are u, v, w and $u \geq v \geq w$. Suppose the distance of u, v on the circumference of the circle be p; the distance of u, w on the circumference is q; the distance of v, w on the circumference is r. Since

$$k^2 - p^2 + k^2 - q^2 = (k + p)(k - p) + (k + q)(k - q)$$
$$> r(k - p) + r(k - q)$$
$$= r(2k - p - q) = r^2,$$

then,

$$S = uvp^2 + uwq^2 + vwr^2 \le uvp^2 + uwq^2 + vw(k^2 - p^2) + vw(k - q^2)$$
$$\le uvp^2 + uwq^2 + uv(k^2 - p^2) + uw(k^2 - q^2)$$
$$= k^2(uv + uw) = k^2 u(v + w) \le \frac{k^2}{4} = \frac{n^2}{16}.$$

Since the value of the sum does not increase in the process of adjustment, then the value of the original sum also does not exceed $\dfrac{n^2}{16}$.   □

Second Day
(8:00 – 12:30; March 10, 2019)

4 Do there exist two positive integer sets A, B that satisfy the following three conditions?

(1) A is a finite set containing at least two elements and B is an infinite set.

(2) There exists positive integer M such that any two different elements in set $S = \{a + b \mid a \in A, b \in B\}$ greater than M are coprime.

(3) For any two coprime positive integers m, n, there exist infinite many $x \in S$ satisfying $x \equiv n(\mod m)$. (Contributed by Yu Hongbing)

Solution   Such positive integer sets do not exist. We will use proof by contradiction. Suppose there exists sets A, B that satisfy the given conditions. Since there are infinite many prime numbers and A is a finite set, we can find prime $p > \max A$. Let $m = \dfrac{(2p - 1)!}{p}$. Then p, m are coprime. By condition (3), there exist infinite many numbers in S that are congruent to p modulo m. By the pigeonhole principle, there exists $a \in A$ and infinite many elements $b \in B$ satisfying $a + b \equiv p(\mod m)$. That is, there are infinite many $b \in B$ such that $b \equiv p - a(\mod m)$. Denote the set composed of these b as $B' \subset B$.

Since A contains at least two elements, take another element $a' \in A, a' \neq a$. For $b \in B'$, there is
$$a' + b \equiv p - a + a' (\bmod\, m).$$
Therefore, S has infinite many elements congruent to $p - a + a'$. modulo m

Since $2 \le p - a + a' \le 2p - 1$ and $p - a + a' \neq p$, then $p - a + a' \mid m$. Thus there are infinite many elements in S that are divisible by $p - a + a'$, and these elements are not mutually prime. This contradicts condition (2). $\qquad\square$

5 As shown in Fig. 5.1, in $\triangle ABC$, $AB > AC$, M is the midpoint of side BC, and BC is the diameter of $\odot M$. Lines AB, AC intersect $\odot M$ at points D (different from B) and E (different from C), respectively. It is known that point P inside $\triangle ABC$ satisfies
$$\angle PAB = \angle ACP, \angle CAP = \angle ABP,$$
$$BC^2 = 2DE \cdot MP.$$
Point X outside $\odot M$ satisfies $XM//AP$ and $\dfrac{XB}{XC} = \dfrac{AB}{AC}$.
Prove that $\angle BXC + \angle BAC = 90°$. (Contributed by Lin Tianqi)

Solution As shown in Fig. 5.2, let the circumscribed circle of $\triangle PBC$ be Γ. The extension of AP intersects Γ at point N. Then
$$\angle BPN = \angle BAP + \angle ABP$$
$$= \angle BAP + \angle CAP = \angle BAC.$$
Similarly, we have $\angle CPN = \angle BAC$. Thus PN bisects $\angle BPC$ and N is the midpoint of arc $\overset{\frown}{BNC}$ of Γ.

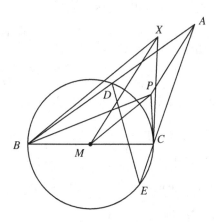

Fig. 5.1

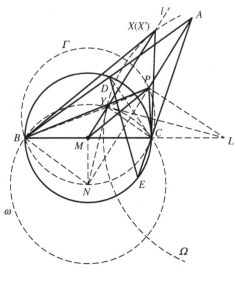

Fig. 5.2

Make circle ω with N as its center and NB as its radius. Make ray l with M as its initial point and $\overrightarrow{PA}$ as the direction. Suppose l and ω intersect at the point Y. By the condition it is easy to know that $CD\perp AD$ and $\triangle ABC \sim \triangle AED$. Thus $\dfrac{DE}{BC} = \dfrac{AD}{AC} = \cos\theta$, where $\theta = \angle BAC$. From the condition, we have

$$PM = \frac{BC^2}{2DE} = \frac{BC}{2\cos\theta} = \frac{BM}{\cos\angle NBC} = NB = NY.$$

Noting that $YM//PN$, we see that $PYMN$ is an isosceles trapezoid, and thus the four points P, Y, M, N are concyclic.

Suppose the exterior angle bisector of $\angle BPC$ intersects the extension of BC at the point L. It is easy to find $PL\perp PN, ML\perp MN$, and thus L, P, M, N are concyclic. Therefore, L, P, Y, M, N are concyclic. Consequently,

$$\angle LYN = \angle LMN = 90°.$$ That is, LY is the tangent to ω at point Y. Therefore, $\triangle LBY \sim \triangle LMN$, so $\dfrac{YB}{YC} = \dfrac{LB}{LY} = \dfrac{LY}{LC}$.

By the condition it is easy to get $\triangle ABP \sim \triangle CAP$, and thus $\dfrac{AB}{AC} = \dfrac{PB}{PA} = \dfrac{PA}{PC}$. Since PL is the exterior angle bisector of $\angle BPC$, then

$\dfrac{LB}{LC} = \dfrac{PB}{PC}$. Therefore,

$$\frac{YB}{YC} = \sqrt{\frac{LB}{LY} \cdot \frac{LY}{LC}} = \sqrt{\frac{LB}{LC}} = \sqrt{\frac{PB}{PC}} = \sqrt{\frac{PB}{PA} \cdot \frac{PA}{PC}} = \frac{AB}{AC}.$$

Take point X' on ray l that satisfies $MX' \cdot MY = MB^2 = MC$. In the following we will prove that $X' = X$.

From the way X' is taken, we know that $\triangle MX'B \sim \triangle MBY$, $\triangle MX'C \sim \triangle MCY$. Therefore,

$$\frac{X'B}{YB} = \frac{BM}{MY} = \frac{CM}{MY} = \frac{X'C}{YC},$$

namely, $\dfrac{X'B}{X'C} = \dfrac{YB}{YC} = \dfrac{AB}{AC}$.

Let Ω denote the trajectory of point T in the plane satisfying $\dfrac{TB}{TC} = \lambda = \dfrac{AB}{AC}$. Then Ω is an Apollonius circle about B, C and λ. It is easy to see that Ω intersects at most two points with ray l. From the method of taking the points Y, X' and the above proof, we see that Y, X' are the intersection of Ω and l. And since

$$\angle BYC = 180° - \frac{1}{2}\angle BNC = 180° - \frac{1}{2}(180° - \angle BPC)$$

$$= 90° + \frac{1}{2}\angle BPC = 90° + \angle BAC,$$

$$\angle BX'C = \angle BX'M + \angle CX'M = \angle YBM + \angle YCM$$

$$= 180° - \angle BYC = 90° + \angle BAC,$$

then Y lies inside $\odot M$ and X' lies outside $\odot M$.

By the given condition, we know that X is also the intersection of Ω, l and lies outside $\odot M$. Therefore, $X = X'$.

Consequently, $\angle BXC + \angle BAC = \angle BX'C + \angle BAC = 90°$. The conclusion has been proved. $\square$

6 Let p, q be two integers that are coprime and greater than 1. We call those positive integers that cannot be expressed in the form $px + qy$ (x, y are non-negative integers) are "(p, q)-bad number". Let $S(p, q)$ be the sum of all the (p, q)-bad numbers to the 2019th power.

Prove that there exists positive integer λ such that for any p, q mentioned above, $(p - 1)(q - 1)$ divides $\lambda S(p, q)$. (Contributed by Wang Bin)

Solution We will prove a more general conclusion: for any positive integer m, there exists positive integer λ_m satisfying that for any two positive integers p, q that are coprime and greater than 1, $(p-1)(q-1)$ divides $\lambda_m S_m(p,q)$. Here $S_m(p,q)$ denotes the sum of all (p,q)-bad numbers to the mth power.

For simplicity, we call "(p,q)-bad number" as "bad number", and call those non-negative integers that are not bad numbers as "good numbers". The following two claims are needed in the proof.

Claim 1 *The largest bad number is $pq - p - q$, and if non-negative integers t_1, t_2 satisfy $t_1 + t_2 = pq - p - q$, then exactly one of them is bad number..*

Claim 2 *For each non-negative integer m, there exists a polynomial with rational coefficients f_m such that for any positive integer n,*

$$f_m(n) = \sum_{i=1}^{n} i^m,$$

with the degree of f_m equal to $m+1$ and the constant term equal to 0.

Let $M = (p-1)(q-1)$. By Claim 1, we know that exactly half of the M elements in set $\{0, 1, 2, \ldots, M-1\}$ are bad numbers, which form set A, and the other half are good numbers, which form set B. We have $|A| = (pq - p - q) - |B|$.

We consider set $T = \{xp - yq \mid x = 1, 2, \ldots, q-1, y = 1, 2, \ldots, p-1\}$. Since all the M elements in T are not equal to each other modulo pq, then $|T| = M$. If $xp - yq > 0$, then it is a bad number. Because if $xp - yq = up + rq, u, v \geq 0$, then $x \equiv u \pmod{q}$; since $1 \leq x \leq q-1$, then $u \geq x$, and thus $v \leq -y < 0$, a contradiction. Similarly, if $yq - xp > 0$, then it is also a bad number. Hence in T, there are exactly $\dfrac{M}{2}$ bad numbers and $\dfrac{M}{2}$ their opposites, i.e., $T = A \cup (-A)$.

Firstly, we prove the conclusion for the case m is even. We have

$$2S_m(p,q) = \sum_{t \in T} t^m = \sum_{x=1}^{q-1} \sum_{y=1}^{p-1} (xp - yq)^m$$

$$= \sum_{k=0}^{m} (-1)^k C_m^k p^k q^{m-k} \left(\sum_{x=1}^{q-1} x^k\right)\left(\sum_{y=1}^{p-1} y^{m-k}\right)$$

$$= \sum_{k=0}^{m} (-1)^k C_m^k p^k q^{m-k} f_k(q-1) f_{m-k}(p-1).$$

By Claim 2, there exists positive integer μ_m such that $\mu_m f_m$ is a polynomial with integer coefficients and the zero constant term. Therefore, for any positive integer n, there is $n \mid \mu_m f_m(n)$. Take $\lambda_m = 2(\mu_0 \mu_1 \ldots \mu_m)^2$. Since

$$(p-1)(q-1) \mid \lambda_m f_k(q-1) f_{m-k}(p-1), k = 0, 1, \ldots, m,$$

we have

$$(p-1)(q-1) \mid \lambda_m S_m(p,q).$$

Next we consider $m = 2l - 1, l \geq 1$. By induction on l, assume the conclusion holds when l is small (if exists). We have

$$\sum_{b \in B} b^{2l} = \sum_{a \in A} (M - 1 - a)^{2l} \equiv \sum_{a \in A} (a+1)^{2l} = \sum_{k=0}^{2l} \mathrm{C}_{2l}^k S_k(p,q) (\mathrm{mod}\ M).$$

And then

$$f_{2l}(M-1) = \sum_{a \in A} a^{2l} + \sum_{b \in B} b^{2l} = S_{2l}(p,q) + \sum_{k=0}^{2l} \mathrm{C}_{2l}^k S_k(p,q).$$

Therefore,

$$2L S_{2l-1}(p,q) = f_{2l}(M-1) - 2S_{2l}(p,q) - \sum_{k=0}^{2l-2} \mathrm{C}_{2l}^k S_k(p,q)$$

$$\equiv f_{2l}(M) - 2S_{2l}(p,q) - \sum_{k=0}^{2l-2} \mathrm{C}_{2l}^k S_k(p,q) (\mathrm{mod}\ M).$$

By the conclusion when m is even and the induction hypothesis, for $k = 0, 1, \cdots, 2l - 2$ and $k = 2l$ there exist positive integer λ_k, such that $M \mid \lambda_k S_k(p,q)$ and $M \mid \mu_{2l} f_{2l}(M)$. Just taking $\lambda_{2l-1} = 2l_{\mu_2 i} \lambda_0 \lambda_1 \ldots \lambda_{2l-2} \lambda_{2l}$, we then have $M \mid \lambda_{2l-1} S_{2l-1}(p,q)$.

Remark about Claim 1 Suppose p, q are coprime and greater than 1. Each integer n can be uniquely expressed as $n = px + qy$, where $x, y \in \mathbb{Z}, 0 \leq y \leq p - 1$. If $x < 0$, then n cannot be expressed as a combination of non-negative integer coefficients of p, q. When $n \geq (p-1)(q-1)$, since $y \leq p - 1$, then $x \geq 0$. When $n = pq - p - q$, we have $-q \equiv qy(\mathrm{mod}\ p)$, namely, $y = p - 1$. Then $x = -1$, and thus $pq - p - q$ is the largest bad number.

For $0 \leq t \leq pq - p - q$, let $t = px + qy, 0 \leq y \leq p - 1$. Then

$$pq - p - q - t = pq - p - q - px - qy = (-x - 1)p + (p - 1 - y)q.$$

Since $0 \leq p - 1 - y \leq p - 1$ and exactly one of x and $-x - 1$ is less than 0, then exactly one of t and $pq - p - q - t$ is a bad number.

Remark about Claim 2   It is easy to find $f_0(x) = x, f_1(x) = \dfrac{1}{2}x(x+1)$. We assume $m \geq 2$. Let $\mathbb{Q}[x]_m$ be the set consisting of all the polynomials with degree not greater than m and rational coefficients. Then $\mathbb{Q}[x]_m$ is an $m+1$ dimensional linear space on the domain of rational numbers, in which $1, x, x^2, \ldots, x^m$ form a basis. On the other hand, polynomials

$$\mathrm{C}_x^0, \mathrm{C}_x^1, \cdots, \mathrm{C}_x^m$$

also constitute a basis of $\mathbb{Q}[x]_m$, here $\mathrm{C}_x^0 = 1$, $\mathrm{C}_x^i = \dfrac{x(x-1)\cdots(x-i+1)}{i!}$, for $i \geq 1$. Thus there exist rational numbers $c_0, c_1, \ldots, c_m$ such that

$$x^m = c_0\mathrm{C}_x^0 + c_1\mathrm{C}_x^1 + \cdots + c_m\mathrm{C}_x^m.$$

It is clear that $c_m \neq 0$. By substituting $x = 0$, we get $c_0 = 0$, and thus

$$x^m = c_1\mathrm{C}_x^1 + c_2\mathrm{C}_x^2 + \cdots + c_m\mathrm{C}_x^m.$$

Let $x = 1, 2, \ldots, n$, and sum them up, by using the following equations, for $j \geq 1$

$$\mathrm{C}_1^j + \mathrm{C}_2^j + \mathrm{C}_3^j + \cdots + \mathrm{C}_n^j = \mathrm{C}_2^{j+1} + \mathrm{C}_2^j + \mathrm{C}_3^j + \cdots + \mathrm{C}_n^j$$
$$= \mathrm{C}_3^{j+1} + \mathrm{C}_3^j + \cdots + \mathrm{C}_n^j = \cdots = \mathrm{C}_{n+1}^{j+1}.$$

Then we have

$$\sum_{i=1}^n i^m = \sum_{j=1}^m \left(c_j \sum_{i=1}^n \mathrm{C}_i^j\right) = \sum_{j=1}^m c_j\mathrm{C}_{n+1}^{j+1}.$$

Therefore,

$$f_m(x) = \sum_{j=1}^m c_j\mathrm{C}_{x+1}^{j+1}.$$

From the above equation, it is easy to find that $f_m(x)$ is a polynomial of degree $m+1$ with rational coefficients and the zero constant term. $\square$

Test III, First Day
(8:00 – 12:30; March 22, 2019)

1 Suppose complex numbers x, y, z satisfy $|x|^2 + |y|^2 + |z|^2 = 1$. Prove that

$$\left|x^3 + y^3 + z^3 - 3xyz\right| \leq 1.$$

(Contributed by Liu Zhipeng)

Solution Let $\omega = e^{\frac{2\pi i}{3}}$. By factorization, we have

$$x^3 + y^3 + z^3 - 3xyz = (x + y + z)(x + \omega y + \omega^2 z)(x + \omega^2 y + \omega z).$$

Therefore,

$$\left| x^3 + y^3 + z^3 - 3xyz \right|^2$$

$$= |x + y + z|^2 \left| x + \omega y + \omega^2 z \right|^2 \left| x + \omega^2 y + \omega z \right|^2$$

$$\leq \frac{1}{27} (|x + y + z|^2 + \left| x + \omega y + \omega^2 z \right|^2 + \left| x + \omega^2 y + \omega z \right|^2)^3$$

$$= \frac{1}{27} ((x + y + z)(\bar{x} + \bar{y} + \bar{z}) + (x + \omega y + \omega^2 z)(\bar{x} + \omega^2 \bar{y} + \omega \bar{z})$$

$$+ (x + \omega^2 y + \omega z)(\bar{x} + \omega \bar{y} + \omega^2 \bar{z}))^3$$

$$= (|x|^2 + |y|^2 + |z|^2) = 1.$$

Consequently, $\left| x^3 + y^3 + z^3 - 3xyz \right| \leq 1$. $\qquad\square$

2 Let set S consisting of positive integers satisfy the following conditions.

(1) $1 \in S$;
(2) for any integer $n > 1$, $n \in S$ if and only if

$$\sum_{\substack{d|n, d<n \\ d \in S}} d \leq n.$$

If $n \in S$ such that the above equal sign holds, then n is called a "good number". Determine the necessary and sufficient condition for positive integer m and odd prime p such that $2^m p$ is a good number. (Contributed by Yu Hongbing)

Solution We will first prove the following lemma.

Lemma *Let p be an odd prime number and integer $r \geq 2$ satisfying $2^r - 1 \leq p < 2^{r+1} - 1$. Then for positive integer k, there is*

$$2^k p \in S \Leftrightarrow r \nmid k.$$

Proof of the lemma It is clear that $p \in S$ and it is easy to prove by induction that $2^l \in S$ for $l \geq 0$. Suppose $k > 0$, and for positive integer

$l < k$, $2^l p \in S$ if and only if $r \nmid l$. Let $k = \alpha r + \beta$, where $\alpha \geq 0, 1 \leq \beta \leq r$. Then

$$\sum_{\substack{d \mid 2^k p, d < 2^k p \\ d \in S}} d = 1 + 2 + \cdots + 2^k + \sum_{\substack{0 \leq j \leq k-1 \\ j=0 \text{ or } r \nmid j}} 2^j p$$

$$= 2^{k+1} - 1 + p\left(2^k - \sum_{j=0}^{\alpha} 2^{jr}\right) \tag{1}$$

$$= 2^{k+1} - 1 + 2^k p - p \cdot \frac{2^{(\alpha+1)r} - 1}{2^r - 1}$$

If $r \nmid k$, namely, $1 \leq \beta < r$, then by $p \geq 2^r - 1$ and $(\alpha + 1)r \geq k + 1$, we know that the right of Formula ①,

$$2^{k+1} - 1 + 2^k p - p \cdot \frac{2^{(\alpha+1)r} - 1}{2^r - 1} \tag{2}$$
$$\leq 2^{k+1} - 1 - (2^{(\alpha+1)r} - 1) + 2^k p \leq 2^k p,$$

and thus $2^k p \in S$.

If $r \mid k$, namely, $\beta = r$, then by $p \leq 2^{r+1} - 2$, we know that the right of Formula ①

$$2^{k+1} - 1 + 2^k p - p \cdot \frac{2^{(a+1)r} - 1}{2^r - 1}$$

$$\geq 2^{k+1} - 1 - 2(2^k - 1) + 2^k p = 2^k p + 1 > 2^k p,$$

and thus $2^k p \notin S$.

In the following we will prove that $2^m p$ is a good number if and only if $p = 2^r - 1$ is a Mersenne prime and

$$m \equiv -1 (\mathrm{mod}\, r).$$

We assume $2^m p$ is a good number. Let $2^r - 1 \leq p < 2^{r+1} - 1$. The other symbols are as in the lemma.

By the lemma, $2^m p \in S \Rightarrow r \nmid m$. Since $2^m p$ is a good number, then Formula ② must take the equal sign, and thus $p = 2^r - 1, (\alpha + 1)r = k + 1$, namely, $\beta = r - 1$. Therefore, $m \equiv -1 (\mathrm{mod}\, r)$.

Conversely, if $p = 2^r - 1$ and $m \equiv -1 (\mathrm{mod}\, r)$, then $r \nmid m$. By the lemma, we have $2^m p \in S$ and Formula ② takes the equal sign. Therefore, $2^m p$ is a good number. $\qquad \square$

3 Let $\mathbb{N}^*$ be the set of all the positive integers. Does there exist a one-to-one map $f : \mathbb{N}^* \to \mathbb{N}^*$ and positive integer k such that each positive integer can be colored in one of the k colors, satisfying that for any positive integers $x \neq y$, the colors of the numbers $f(x) + y$ and $x + f(y)$ are different? (Contributed by Xiong Bin)

Solution There exists such a kind of map.

Let $a_i = 4^i, i = 1, 2, \cdots$. After excluding the positive integer powers of 4 from positive integers, the remaining numbers in $\mathbb{N}^*$ are arranged in ascending order $b_1 < b_2 < \cdots$. Then $\{a_1, a_2, \cdots\}$ and $\{b_1, b_2, \cdots\}$ form a partition of the set of positive integers.

The definition of $f : \mathbb{N}^* \to \mathbb{N}^*$ is as follows: for $i \geq 1$, $f(a_i) = b_i$, $f(b_i) = a_i$. It is clear that f is a one-to-one map and is a involution. That is, for $x \in \mathbb{N}^*$, there is $f(f(x)) = x$.

We show that all of the positive integers can be two-colored such that the color of $f(x) + y$ is different from that of $x + f(y)$ for any $x \neq y$.

First, note that for any $x \neq y$, there is $f(x) + y \neq x + f(y)$. Otherwise, we have $f(x) - x = f(y) - y$. We may assume $f(x) = a_i, x = b_i$, $f(y) = a_j, y = b_j, i < j$. Since $b_{n+1} - b_n \leq 2$ and $a_{n+1} - a_n \geq 3$, thus $a_n - b_n$ is strictly monotonic increasing with respect to n. Therefore, $a_j - b_j > a_i - b_i$, a contradiction.

Regarding $\mathbb{N}^*$ a set of vertices, we construct a set of edges as follows: For any different positive integers u, v, if there exist different positive integers x, y such that $f(x) + y = u, f(y) + x = v$, then connect a edge between u and v. Then we obtain a graph $G = (\mathbb{N}^*, E)$. We will show that G does not contain any circle, so the chromatic number of G is 2. Thus the vertices can be two-colored such that adjacent vertices are of different colors. This coloring satisfies the requirements.

We will prove that for any positive integer u, v, there is at most one vertex smaller than u among all the neighbor vertices of u in G. This means that G does not contain any circle. Because if there is a circle, we consider the largest number u on the circle. Since u has two neighbors in the circle, then there are two neighboring numbers that are less than u, a contradiction.

Let u be any positive integer, and consider its neighbor v, $v < u$. There exist two different positive integers x, y such that $u = f(x) + y$, $v = f(y) + x$. Let

$$\{f(x), x\} = \{a_i, b_i\}, \{f(y), y\} = \{a_j, b_j\}.$$

In the following we discuss four cases.

Case 1: $f(x) = a_i, x = b_i, f(y) = a_j, y = b_j$.

Then $u = a_i + b_j$, $v = a_j + b_i$. Since $u > v$, then $a_i + b_j > a_j + b_i$, namely, $a_i - b_i > a_j - b_j$, and thus $i > j$. Therefore, a_i is the largest power of 4 that is not greater than u, and $u - a_i = b_j$ is not a positive integer power of 4. Hence, in this case, u uniquely determines a_i, b_j, and also uniquely determined $v < u$. Therefore, there exists at most one $v < u$ such that u, v are neighboring.

Case 2: $f(x) = a_i, x = b_i, f(y) = b_j, y = a_j$.

Then $u = a_i + a_j, v = b_i + b_j$. It is clear that $u > v$. In this case u is the sum of two positive integer powers of 4, one of them being the largest power of 4 not greater than u. The expression is unique. Therefore, there exists at most one $v < u$ such that u, v are neighboring. Note that Case 1 and Case 2 are mutually exclusive.

Case 3: $f(x) = b_i, x = a_i, f(y) = a_j, y = b_j$.

Then $u = b_i + b_j, v = a_i + a_j$. At this time, $v > u$, contradicting the assumption that $v < u$.

Case 4: $f(x) = b_i, x = a_i, f(y) = b_j, y = a_j$.

Then $u = b_i + a_j, v = a_i + b_j$. This is the same as in Case 1, a_j is the largest power of 4 not greater than u. Then yield the same v (if it exists) as in Case 1.

In summary, for each positive integer u, there exists at most one positive integer v smaller than u that is neighboring u. The conclusion is proved. $\square$

Second Day
(8:00 – 12:30; March 23, 2019)

4 Let $\mathbb{R}$ be the set of all real numbers. Find all functions of two variables $f : \mathbb{R} \times \mathbb{R} \to \mathbb{R}$ satisfying the following conditions:

(1) For any $x, y \in \mathbb{R}$, $f(x, y) = f(y, x)$;

(2) for any $x, y, z \in \mathbb{R}$,

$$(f(x,y) - f(x,z))(f(x,y) - f(y,z))(f(x,z) - f(y,z)) = 0;$$

(3) for any $x, y, a \in \mathbb{R}$,

$$f(x + a, y + a) = f(x, y) + a;$$

(4) $f(0, x)$, as a function of x, is monotonic non-decreasing on $\mathbb{R}$.

(Contributed by Lin Bo)

Solution For any real number c, if f satisfies Conditions (1)–(4), then $f(x, y) - c$ also satisfies them. Hence only the case of $f(0, 0) = 0$ needs to be considered. By conditions (3) and (4), it follows that when x is fixed $f(x, y)$ is monotonic non-decreasing with respect to y, and then by symmetry, $f(x, y)$ is monotonic non-decreasing about x when y is fixed. And by Condition (3), $f(x, x) = x + f(0, 0) = x$. Then for $x < y$, there is

$$x = f(x, x) \leq f(x, y) \leq f(y, y) = y.$$

We will find f in several steps.

(i) Let $a > 0$. If $f(0, a) = 0$ or $f(0, -a) = -a$, then for any $x \in [0, a]$, there is $f(0, x) = 0$, $f(0, -x) = -x$. If $f(0, a) = a$ or $f(0, -a) = 0$, then for any $x \in [0, a]$, there is $f(0, x) = x$, $f(0, -x) = 0$.

We prove the case $a = 0$. The other cases can be similarly proved. If $f(0, a) = 0$, then by condition (4) we know that for $x \in [0, a]$, there is $0 = f(0, 0) \leq f(0, x) \leq f(0, a) = 0$, and thus $f(0, x) = 0$. By condition (3), there is $0 = f(0, x) = x + f(-x, 0)$, and hence $f(0, -x) = f(-x, 0) = -x$.

(ii) If there exists $a > 0$ such that $f(0, a) = 0$, then for any $x > 0$ there is $f(0, x) = 0$.

We will prove $f(0, ka) = 0$ by induction on positive integer k. Then reach the conclusion by the monotonicity in Condition (4). It is obvious that the conclusion holds when $k = 1$. Assume the conclusion holds for k but not for $k + 1$. Then

$$f(0, (k + 1)a) > 0.$$

We consider

$$f(0, ka) = 0, f(0, (k + 1)a) > 0,$$

$$f(ka, (k + 1)a) = ka + f(0, a) = ka > 0.$$

By Condition (2) two of the three numbers above are equal, and it can only be $f(0, (k + 1)a) = ka$. For any $x \in (ka, (k + 1)a)$, we consider

$$f(0, x), f(0, ka) = 0, f(ka, x) = ka + f(0, x - ka) = ka.$$

By condition (2), two of the three numbers above are equal. Thus $f(0, x) = 0$ or ka. Combining this with the monotonicity, we know that there exists a unique real number $t \in [ka, (k + 1)a]$ such that for $ka \leq x < t$, there is $f(0, x) = 0$; and for $t < x \leq (k + 1)a$, there is $f(0, x) = ka$. Suppose

$ka \le y < z \le (k+1)a$ satisfies

$$f(0,y) = 0, f(0,z) = ka.$$

As

$$f(y,z) = y + f(0, z - y) = y,$$

by Condition (2) there must be $y = ka$. Then $t = ka$, i.e., for any $x \in (ka, (k+1)a]$, there is $f(0,x) = ka$.

For any $x \in (0, a)$, there is $f(0, x) = 0$, $f(0, (k+1)a) = ka$,

$$f(x, (k+1)a) = x + f(0, (k+1)a - x) = x + ka.$$

The above three numbers are different from each other and are in contradiction with condition (2). Therefore, the conclusion holds for $k + 1$. The proof has been completed.

Combining with the conclusion in Step (i), we know that if there exists $a > 0$ such that $f(0, a) = 0$, then for any $x > 0$, $f(0, x) = 0$, $f(0, -x) = -x$, i.e., $f(0, x) = \min\{0, x\}$. Then combining with Condition (3), we have

$$f(x, y) = \min\{x, y\}.$$

In the following we shall assume Condition (5): for any $x > 0$, $f(0, x) > 0$.

(iii) $f(0, x)$ has no upper bound for $x > 0$.

We use proof by contradiction. Suppose there is an upper bound and take its supremum $M > 0$. If there exists $x > 0$ such that $f(0, x) = b < M$, we can take $y > 0$ such that $f(0, y) > \max\{M - x, b\}$ because M is the supremum. At this point we have

$$M \ge f(0, x + y) \ge f(0, y) > b,$$

$$f(x, x + y) = x + f(0, y) > M.$$

Then $f(0, x) = b, f(0, x + y), f(x, x + y)$ are different from each other. This is in contradiction with Condition (2).

Therefore, for any $x > 0$, there is $f(0, x) = M$. In particular, $f(0, M) = M$. But by the conclusion in Step (i), for $x \in (0, M)$, $f(0, x) = x$. A contradiction.

(iv) Set $\{x > 0 \mid f(0, x) = x\}$ has no upper bound.

We use proof by contradiction. Assume there exists $N > 0$ such that for any $x > N$, there is $0 < f(0, x) < x$. By the conclusion in Step (iii), we can take $x > N$ such that $f(0, x) > N$. Let $y = f(0, x)$. Then $N < y < x$, and $f(0, y) < y$, $f(0, x) = y$, $f(y, x) = y + f(0, x - y) > y$ are different from each other, contradicting Condition (2).

By the conclusion in Step (iv), there exists a sequence of positive real numbers $\{x_n\}$ that tends to infinity and satisfies $f(x_n) = x_n$. Then by the conclusion in Step (i), for any $x > 0$, there is $f(0, x) = x, f(0, -x) = 0$, i.e. $f(0, x) = \max\{0, x\}$. Then by condition (3), we know that $f(x, y) = \max\{x, y\}$.

In summary, the function f satisfying the required conditions can only be $f(x, y) = \min\{x, y\} + c$ or $f(x, y) = \max\{x, y\} + c$, where c is a constant. It is easy to verify that these two types of functions do satisfy Conditions (1)–(4). $\square$

5 In acute triangle ABC, $AB < AC$. AD is the height on side BC, with foot point D. I is the incenter of $\triangle ABC$, and J is the escenter corresponding to vertex A. Point E is on side AB and point F is on the extension line of AB satisfying

$$BE = BF = BD.$$

Prove that: There exist two points P, Q (allowed to coincide), on the circumscribed circle of $\triangle ABC$ satisfying $PB = QC$ and $\triangle PEI$ is similar to $\triangle QFJ$ with the vertices corresponding in this order. (Contributed by He Yijie)

Solution   As shown in the Fig. 5.1

Let the circumscribed circle of $\triangle ABC$ be ω. Let AJ intersects side BC, arc $\overset{\frown}{BC}$ at points K, M, respectively. Then M is the midpoint of $\overset{\frown}{BC}$.

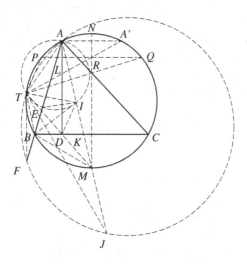

Fig. 5.1

Suppose line DM intersects ω at point T other than M. Since $\overarc{BM} = \overarc{CM}$, then $\triangle DBM = \triangle BTM$, and thus $\triangle DBM \sim \triangle BTM$. By the properties of the incenter, we have $MB = MI$. Hence

$$MI^2 = MB^2 = MD \cdot MT,$$

and then $\triangle DIM \sim \triangle ITM$. Therefore, $\angle MTI = \angle MID$. And since

$$\angle ATM = \angle ABM = \angle ABC + \angle MBC = \angle ABC + \angle BAM = \angle AKC,$$

then

$$\angle ATI = \angle ATM - \angle MTI = \angle AKC - \angle MID = \angle IDK.$$

And since $BD = BE$ and BI bisects $\angle ABC$, then $\triangle BID \cong BIE$, and thus $\angle IDK = \angle IEA$. Therefore, $\angle ATI = \angle AEI$, and T, A, I, E are concyclic. Similarly, we can prove that T, A, J, F are concyclic (In the above proof, replace I with J and E with F). Consequently,

$$\angle TEF = 180° - \angle TEA = 180° - \angle TIA = \angle TIJ.$$

Similarly, we can obtain

$$\angle TFE = \angle TJI,$$

and thus $\angle TFE = \angle TJI$.

By the properties of escenter, we learn that M is the midpoint of IJ and B is the midpoint of EF, and thus $\triangle TEB \sim \triangle TIM$.

Make a parallel line of BC through A, which intersects circle ω at another point A' other than A. Let TA' intersects the diameter MN of circle ω at point R and intersects AD at point L. It is clear that R is the midpoint of $A'L$. Make a parallel line of BC through R and it intersects circle ω at points P and Q. We may set point P on the minor arc $\overarc{AB}$. It is obvious that $BP = CQ$. In the following we will prove that $\triangle PEI \sim \triangle QFJ$.

Firstly we prove $\triangle TPR \sim \triangle TIM$. Since $AA'//PQ$, then

$$\angle TRP = \angle TA'A = \angle TMI.$$

By power of a point theorem and $\triangle DIM \sim \triangle ITM$, we have

$$\left(\frac{PR}{TR}\right)^2 = \frac{PR \cdot RQ}{TR^2} = \frac{TR \cdot RA'}{TR^2} = \frac{RA'}{TR} = \frac{RL}{TR} = \frac{MD}{TM} = \left(\frac{MI}{MT}\right)^2.$$

That is, $\dfrac{PR}{TR} = \dfrac{MI}{MT}$. Therefore, $\triangle TPR \sim \triangle TIM$.

And as R, M are the midpoints of PQ, IJ, respectively, then $\triangle TPQ \sim \triangle TIJ$.

It follows that $\triangle TPQ \sim \triangle TIJ \sim \triangle TEF$, and then $\triangle PEI \sim \triangle QFJ$. $\qquad\square$

6 Let $G = (V, E)$ be a simple graph. $A \subset V$ is a non-empty proper subset, and r, d, l be positive integers, $r \geq 3, 2d \leq l < rd$. Suppose $f : V \to \{1, 2, \cdots, l\}$. It is known that:

(1) For any edge $uv \in E$, there is $d \leq |f(u) - f(v)| \leq l - d$;

(2) for any normal coloring with r colors to the vertices of G, the vertices of the same color will be either all or none in A.

Prove that there exists a normal coloring with $r - 1$ colors for the vertices of G.

Remark A coloring to the vertices in G is said to be normal, if any two adjacent vertices are colored with different colors. (Contributed by Qu Zhenhua)

Solution Take a circle ω with circumference l. The arc distance between two points x, y on the circumference is the length of the minor arc between x, y, denoted as $L(x, y)$. Let l points uniformly distribute on ω, labeled by $1, 2, \cdots, l$, forming set S. We can equivalently treat S as $\{1, 2, \ldots, l\}$, and consider $f : V \to S$. Then Condition (1) is equivalent to the fact that for any $uv \in E$, there is $L(f(u), f(v)) \geq d$. Condition (1) has rotation and symmetry invariance for $f(x)$: that means adding an integer c to $f(x)$, or multiplying $f(x)$ by -1 (both in the sense of modulo l), Condition (1) is still satisfied.

We will first prove that, for any $u \in A, v \in V \backslash A$, there is $L(f(u), f(v)) \geq d$. We use proof by contradiction. Suppose there exist $u \in A, v \in V \backslash A$ such that $L(f(u), f(v)) \leq d - 1$. By rotation and symmetry, we may set $f(u) = 1, 1 \leq f(v) \leq d$. Then a normal coloring of G with r colors can be done as follows: If $f(x) \in [(i - 1)d, id)$, $i \geq 1$, we can color x with color i. Since for any $xy \in E$, there is $L(f(x), f(y)) \geq d$, thus it is impossible for both $f(x)$ and $f(y) \in [(i - 1)d, id)$. Therefore, x, y are with different colors, and this is normal coloring. As $l < rd$, at most r colors are used. However, $f(u), f(v) \in [1, d)$, and u, v are colored with the same color, which contradicts Condition (2).

In the following, we shall prove the original proposition. Since both A and $V \backslash A$ are non-empty sets, then let $u \in A$ and $v \in V \backslash A$ such that $L(f(u), f(v))$ takes the minimum value. By the previous proof,

$L(f(u), f(v)) = p \geq d$. By translation and symmetry, we may set $f(u) = p, f(v) = l$. And by the minimum, there does not exist $x \in V$ such that $f(x) \in [1, p)$. Otherwise, if $x \in A$, $L(f(x), f(v)) < L(f(u), f(v))$; if $x \in V \backslash A$, $f(f(u), f(x)) < L(f(u), f(v))$ — both contradict the minimum of $L(f(u), f(v))$.

Now we can do the following normal coloring of G with $r - 1$ colors: If $f(x) \in [id, (i + 1)d), i \geq 1$, then x will be colored with color i. Since there does not exist $x \in V$ such that $f(x) \in [1, d)$, then each vertex is colored with a color. In the same way as the previous coloring method, we see that this is a normal coloring. Since $f(x) \leq l < rd$, we know that at most $r - 1$ colors are used. In this way, the $r - 1$ color normal coloring of G was obtained. The conclusion is confirmed. □

Test IV, First Day
(8:00 – 12:30; March 26, 2019)

1 Quadrilateral $ABCD$ is inscribed in $\odot O$. A line through O intersects sides AB, AD at points E and F, respectively, with $OE = OF$. Points M, N are the midpoints of sides BC, CD, respectively. Lines EN, FM intersect at point S. Line OS intersects the lines of sides AB, AD at two different points P, Q. Suppose the excentre of $\triangle EFS$ is T.

Prove that: if quadrilateral $OTSC$ is a parallelogram, then $AP = AQ$. (Contributed by Lin Tianqi)

Solution As shown in Fig. 1.2, suppose the midpoints of SE, SF, SO, ST be E', F', O', T', respectively. Then T' is the excentre of $\triangle SE'F'$. Since

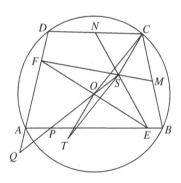

Fig. 1.1

quadrilateral $OTSC$ is a parallelogram, then

$$\overrightarrow{O'T'} = \frac{1}{2}\overrightarrow{OT} = \frac{1}{2}\overrightarrow{CS}.$$

Combing with the properties of orthocenter, we learn that C is the ortho-center of $\triangle SE'F'$, and thus

$$CE' \perp SF', CF' \perp SE'.$$

In addition, since OE' and OF' are both the medians of $\triangle SEF$, then $\angle OE'C = \angle OF'C = 90°$. And by vertical theorem, there is $\angle OMC = \angle ONC = 90°$, and thus M, N, E', F' are all on the circle with OC as its diameter. Since S is the orthocenter of $\triangle CE'F'$, we know that M, S are symmetric about CE' and N, S are symmetric about CF'. Thus $CM = CS = CN$, then $CB = CD$, and thus AC bisects $\angle BAD$. And since CE' bisects MS and E' is the midpoint of SE, then $CE'//ME$, and hence $\angle ENF = 90°$. Similarly, we can obtain $\angle ENF = 90°$, and thus E, F, N, M are concyclic.

Construct the symmetry point L of point T about line EF. Noting that $\triangle EFS \sim \triangle MNS$ and that T, C are their excentres, respectively. Then

$$\angle ELF = \angle ETF = \angle MCN = 180° - \angle EAF.$$

Therefore, A, E, L, F are concyclic. As $LE = LF$, then AL bisects $\angle EAF$, and hence A, L, C are collinear. And because $\overrightarrow{OL} = \overrightarrow{TO} = \overrightarrow{SC}$, so $OSCL$ is a parallelogram, then $OS//LC$, namely, $OS//AC$. Since AC bisects $\angle BAD$, then

$$\angle APQ = \angle CAB = \angle CAD = \angle AQP.$$

Therefore, $AP = AQ$. $\qquad\square$

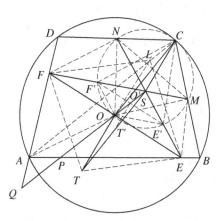

Fig. 1.2

2 Let $G = (V, E)$ be a simple graph of order 2019. It is known that

(1) G contains no triangles, but adding an edge between any two non-adjacent vertices will produce triangles;

(2) $|E| \geq 2019$.

Find the minimum possible value of $|E|$. (Contributed by Qu Zhenhua)

Solution We consider the following example. The 2019 vertices are denoted as $A, B, C, D, P_1, \ldots, P_{2015}$, and the edges are

$$AB, BC, CD, AP_i, DP_i, 1 \leq i \leq 2015.$$

The graph does not contain triangles, but adding edges between any two non-adjacent vertices gives triangles. Its edge number is $E = 3 + 2 \times 2015 = 4033$.

In the following we shall show that this is the minimum. Suppose graph $G = (V, E)$ satisfies the conditions in the problem. We discuss the minimum degree δ of G in the following.

(i) $\delta = 0$. Then there is a vertex u not adjacent to any other vertex. Take any vertex $v \neq u$. After adding edge uv, the graph still does not contain a triangle, which contradicts Condition (1).

(ii) $\delta = 1$. Then there is a vertex u adjacent to another vertex v only. For any $w \neq u, v$, if $vw \notin E$, the graph still does not contain a triangle after adding edge uw. Therefore, $vw \in E$. Then v is adjacent to all other vertices, and there are no more other edges. Thus this is a star graph $K_{1,2018}$ with edge number 2018, Contradicting Condition 2.

(iii) $\delta = 2$. Suppose u is adjacent to v, w only. For any $x \neq u, v, w$, and not adjacent to u, by Condition (1), there exists another vertex which is adjacent to both u and x. This vertex can only be v or w, so x is adjacent to at least one of v, w. Then $V \backslash \{v, w\}$ can be divided into three classes V_1, V_2, V_3, each containing vertices adjacent to v only, vertices adjacent to w only, and vertices adjacent to both v and w, respectively. $u \in V_3$.

By the fact that G does not contain triangles, we know that the vertices in $V_1 \cup V_3$ are not adjacent to each other, and neither do the vertices in $V_2 \cup V_3$. For any $x \in V_1, y \in V_2$, x and y are adjacent, otherwise, there exists no vertex that is adjacent to both x and y, and adding edge xy will not produce any triangle, contradicting to Condition (1).

Case 1: $V_1 = \varnothing$ or $V_2 = \varnothing$. We may let $V_1 = \varnothing$. Since the vertices in $V_2 \cup V_3$ are not adjacent to each other, then G is a bipartite graph with one part of vertices being $\{v, w\}$ and the other part of vertices being $V\backslash\{v, w\}$. If it is not a complete bipartite graph, then an edge can be added between the two parts and it will still be a bipartite graph, containing no triangle, which contradicts condition (2). Therefore, G is a complete bipartite graph $K_{2,2017}$ whose edge number is 4034.

Case 2: V_1, V_2 are non-empty. Let $|V_1| = a, |V_2| = b, |V_3| = 2017 - a - b$, $a, b \geq 1$. Then

$$|E| \geq 2(2017 - a - b) + a + b + ab = (a - 1)(b - 1) + 4033 \geq 4033.$$

(iv) $\delta = 3$. Suppose the degree of u is 3 and it is adjacent to x, y, z. For any $v \neq u, x, y, z$, using similar discussion in (iii) we know that u is adjacent to at least one of the x, y, z vertices. Therefore,

$$\deg(x) + \deg(y) + \deg(z) \geq 2019 - 4 + 3 = 2018.$$

By the relationship between degree and edge, we have

$$|E| = \frac{1}{2}(\deg(x) + \deg(y) + \deg(z) + \sum_{v \neq x, y, z} \deg(v))$$

$$\geq \frac{1}{2}(2018 + 3 \cdot 2016) = 4033.$$

(v) $\delta \geq 4$. Then $|E| \geq \frac{1}{2}\delta|V| \geq 2 \times 2019 = 4038$.

In summary, the minimum of the edge number of G is 4033. $\qquad\square$

③ Given 60 points on the plane, and any three of them are not collinear. Prove that these 60 points can be divided into 20 groups of three points $\{A_i, B_i, C_i\}$, $i = 1, 2, \cdots, 20$ such that

$$\bigcap_{i=1}^{20} \triangle A_i B_i C_i \neq \varnothing,$$

where $\triangle A_i B_i C_i$ denotes the set of all the points in the interior and on the boundary of triangle $A_i B_i C_i$ on the plane. (Contributed by Yao Yijun)

Solution Denote the set of these 60 points by S. We consider the convex hulls $\triangle_1, \triangle_2, \ldots, \triangle_N$ of the C_{60}^{41} subsets, each consisting of 41 elements, where $N = C_{60}^{41}$. We firstly show that

$$\triangle_1 \cap \triangle_2 \cap \cdots \cap \triangle_N \neq \varnothing.$$

According to Helly's theorem for convex sets, we just need to prove that the intersection of any three of them is non-empty.

Suppose $\triangle_i, \triangle_j, \triangle_k$ are the convex hulls of 41-element subsets A, B, C, respectively. Since

$$|A \cap B| = |A| + |B| - |A \cup B| \geq 41 + 41 - 60 = 22,$$

then $|A \cap B| + |C| > 60$, and thus $A \cap B \cap C \neq \varnothing$. Since $A \subset \triangle_i, B \subset \triangle_j, C \subset \triangle_k$, then the intersection of $\triangle_i, \triangle_j, \triangle_k$ is non-empty.

Suppose P is a point on the plane and P is contained in the closure of all the 41-element subsets. Make a ray l with P as the initial point and rotate l counterclockwisely around P, passing through every point in S, which is denoted in order as $P_1, P_2, \ldots, P_{60}$. If at some moment l passes through two points simultaneously, then these two points are arbitrarily noted as P_i, P_{i+1}. If P is also a point in S, then denote $P_{60} = P$.

Take 20 triangles $\triangle P_i P_{i+20} P_{i+40}, i = 1, 2, \ldots, 20$. We will show $P \in P_i P_{i+20} P_{i+40}$. Then the intersection of these 20 triangles is non-empty.

We use proof by contradiction. Suppose $P \notin \triangle P_i P_{i+20} P_{i+40}$. Then a line L can be made through P such that P_i, P_{i+20}, P_{i+40} are all on one side of L. At this point, when l rotates in counterclockwise direction from PP_i to PP_{i+40}, the points passed are on the same side of L. Therefore, the convex hulls of the 41 points $P_i, P_{i+1}, \ldots, P_{i+40}$ is on one side of L and does not contain P, contradicting the selection of P. The conclusion is then confirmed. $\qquad\square$

Second Day
(8:00 – 12:30; March 27, 2019)

4 For any non-empty number set X, denote $S(X)$ as the sum of all the elements in X. Let n be an integer with $n \geq 2$. Prove that: there exist n-element subset T of $\{1, 2, 3, \ldots, 2^n\}$ satisfying that $S(A)$ does not divide $S(B)$ for any two distinct non-empty subsets A, B of T. (Contributed by Yu Hongbing)

Solution 1 We prove that $T = \{2^n - 2^0, 2^n - 2^1, \cdots, 2^n - 2^{n-1}\}$ satisfies the conditions of the problem.

We use proof by contradiction. Assume there exist two different non-empty subsets A, B such that $S(A)$ divides $S(B)$. Suppose

$$A = \{2^n - 2^{a_1}, \ldots, 2^n - 2^{a_p}\}, B = \{2^n - 2^{b_1}, \ldots, 2^n - 2^{b_q}\},$$

where $\{a_1, \ldots, a_p\}$ and $\{b_1, \ldots, b_q\}$ are two different non-empty subsets of $\{0, 1, \ldots, n-1\}$. Then

$$S(A) = 2^n p - \sum_{i=1}^{p} 2^{a_i}, \ S(B) = 2^n q - \sum_{j=1}^{q} 2^{b_b}.$$

Let $S(B) = kS(A), k \geq 1$. Denote $P = \sum_{i=1}^{p} 2^{a_i}, Q = \sum_{j=1}^{q} 2^{b_j}$. Then

$$(kp - q)2^n = kP - Q.$$

If $k = 1$, since $0 < P < 2^n, 0 < Q < 2^n$, then $-2^n < P - Q < 2^n$. As $2^n \mid P - Q$, we have $P = Q$. From the uniqueness of the binary representation, it follows that

$$\{a_1, \cdots, a_p\} = \{b_1, \cdots, b_q\},$$

and then $A = B$. This is inconsistent with the assumption. Therefore, $k \geq 2$. Let $kp - q = t, kP - Q = 2^n t$. Note that p, q are the number of terms in binary representation of P, Q, respectively, so

$$p = P - \sum_{i=1}^{\infty} \left[\frac{P}{2^i}\right] = P - \sum_{i=1}^{n-1} \left[\frac{P}{2^i}\right],$$

$$q = Q - \sum_{i=1}^{n-1} \left[\frac{Q}{2^i}\right].$$

Hence

$$\sum_{i=1}^{n} \left[\frac{kP}{2^i}\right] = \sum_{i=1}^{n} \left[\frac{Q + 2^n t}{2^i}\right]$$

$$= (2^n - 1)t + \sum_{i=1}^{n-1} \left[\frac{Q}{2^i}\right] = (2^n - 1)t + Q - q$$

$$= (2^n - 1)t + kP - 2^n t - q = kP - t - q$$

$$= kP - kp = \sum_{i=1}^{n} k \left[\frac{P}{2^i}\right].$$

Since $k[x] \geq [kx]$ for real number x, the equality sign holds if and only if $\{x\} < \frac{1}{k}$. Therefore, from the above equation we know that

$$k \left[\frac{P}{2^i}\right] = \left[\frac{kP}{2^i}\right], \ 1 \leq i \leq n,$$

and thus

$$\left\{\frac{P}{2^i}\right\} < \frac{1}{k} \le \frac{1}{2}, 1 \le i \le n.$$

However, let $v_2(P) = \alpha, 0 \le \alpha \le n - 1$. Then

$$\left\{\frac{P}{2^{\alpha+1}}\right\} = \frac{1}{2},$$

a contradiction. Therefore, $S(B) = kS(A)$ cannot hold. The constructed set T satisfies the requirement.

Solution 2   The construction of set T and related symbols are the same as in Solutiion 1. We will show that $S(B) = kS(A), k \ge 2$ cannot be valid. This equation can be written as

$$2^n q + k \sum_{i=1}^{p} 2^{a_i} = k \cdot 2^n p + \sum_{i=1}^{q} 2^{b_i}.$$

Consider a blackboard, on which we write q "2^n"s and k "2^{a_1}"s, ..., "2^{a_b}"s, respectively. Then there are totally $q + kp$ number terms on the blackboard. The following operation on the blackboard is allowed: erase two equal numbers that are the power of 2 less than 2^n, say 2^r, and write "2^{r+1}". Since $k \ge 2$, the operation can be performed at least once. After one such operation, the number of number terms on the blackboard decreases by 1, so it can be performed only a finite number of times. When the operation cannot be performed any more, the number terms on the board become several "2^n"s and a few other numbers that are different powers of 2 less than 2^n. Since the operation does not change the sum of all the numbers on the blackboard, and by the uniqueness of the binary representation, it is clear that when the operation cannot be performed, it must become the right-hand side of the equation above, i.e., kp "2^n" terms and one "2^{b_1}", $\cdots$, "2^{b_q}" term, each. But the number of number terms on the board at this point is still $kp+q$, not decreased from the begining. It is a contradiction. □

5   Given positive real numbers a, b, c, x, y, z satisfying the following conditions:

(1) $a + b + c = x + y + z$;

(2) $abc = xyz$;

(3) $a > \max\{x, y, z\}$.

Find all integers n such that $a^n + b^n + c^n \ge x^n + y^n + z^n$. (Contributed by Fu Yunhao)

Solution 1 By symmetry, we may assume $x \geq y \geq z$. Since both the conditions and the conclusion are homogeneous with respect to a, b, c, x, y, z, i.e., multiplying by a positive real number λ at the same time does not affect the conditions and the conclusion, a positive real number λ can be chosen such that $a\lambda = (x\lambda)(y\lambda)$. So we can set $a = xy$. Then from Condition (2) we get $z = bc$. Substituting it into Condition (1), we have $xy + b + c = x + y + bc$, and then

$$(x - 1)(y - 1) = (b - 1)(c - 1). \qquad \textcircled{1}$$

By Condition (3), we have $xy > \max(x, y)$, and thus $x > 1, y > 1$. Then by $\textcircled{1}$, it follows that b, c are both either greater or less than 1.

If b, c are both greater than 1, then by $x \geq z = bc, y \geq z = bc$ have

$$x - 1 \geq bc - 1 > b - 1, y - 1 \geq bc - 1 > c - 1,$$

and thus

$$(x - 1)(y - 1) > (b - 1)(c - 1).$$

This contradicts $\textcircled{1}$. Therefore, $b < 1, c < 1$.

For integer n, we define

$$F(n) = (a^n + b^n + c^n) - (x^n + y^n + z^n)$$
$$= (x^n y^n + b^n + c^n) - (x^n - y^n - b^n c^n)$$
$$= (x^n - 1)(y^n - 1) - (b^n - 1)(c^n - 1).$$

When $n \geq 2$, since

$$\frac{(x^n - 1)(y^n - 1)}{(b^n - 1)(c^n - 1)} = \frac{(1 + x + \cdots + x^{n-1})(1 + y + \cdots + y^{n-1})}{(1 + b + \cdots + b^{n-1})(1 + c + \cdots + c^{n-1})} > 1,$$

then $F(n) > 0$.

When $n = 0, 1$, it is easy to find $F(n) = 0$.

When $n < 0$, let $n = -m$. Then

$$F(n) = \frac{(x^{-m} - 1)(y^{-m} - 1)}{(b^{-m} - 1)(c^{-m} - 1)}$$
$$= \frac{x^{-m} y^{-m}(1 - x^m)(1 - y^m)}{b^{-m} c^{-m}(1 - b^m)(1 - c^m)}$$
$$= \frac{x^{-m} y^{-m}(1 + x + \cdots + x^{m-1})(1 + y + \cdots + y^{m-1})}{b^{-m} c^{-m}(1 + b + \cdots + b^{m-1})(1 + c + \cdots + c^{m-1})}$$
$$= \frac{(x^{-1} + x^{-2} + \cdots + x^{-m})(y^{-1} + y^{-2} + \cdots + y^{-m})}{(b^{-1} + b^{-2} + \cdots + b^{-m})(c^{-1} + c^{-2} + \cdots + c^{-m})} < 1.$$

Therefore, $F(n) < 0$

In summary, the set of required n's is that of all the non-negative integers.   □

Solution 2   By Condition (3), we have

$$
\begin{aligned}
0 &< (a - x)(a - y)(a - z) \\
&= a^3 - (x + y + z)a^2 + (xy + yz + zx)a - xyz \\
&= a^3 - (a + b + c)a^2 + (xy + yz + zx)a - abc \\
&= (a - a)(a - b)(a - c) + (xy + yz + zx - ab - bc - ca)a \\
&= (xy + yz + zx - ab - bc - ca)a,
\end{aligned}
$$

and thus $xy + yz + zx > ab + bc + ca$. For any integer n, let

$$
p_n = a^n + b^n + c^n - x^n - y^n - z^n.
$$

Using the method of characteristic roots of recursive series, it follows that for any integer n there is

$$
\begin{aligned}
&p_{n+3} - (x + y + z)p_{n+2} + (xy + yz + zx)p_{n+1} - xyzp_n \\
&= (xy + yz + zx - ab - bc - ca)(a^{n+1} + b^{n+1} + c^{n+1}) > 0.
\end{aligned}
$$

Thus

$$
p_{n+3} - (x + y)p_{n+2} + xyp_{n+1} > z(p_{n+2} - (x + y)p_{n+1} + xyp_n). \quad ②
$$

Since $p_0 = p_1 = 0$,

$$
p_2 = (a + b + c)^2 - (x + y + z)^2 + 2(xy + yz + zx - ab - bc - ca) > 0,
$$

then $p_2 - (x + y)p_1 + xyp_0 > 0$. Combining with ②, it can be shown by mathematical induction that for $n \geq 0$, we have

$$
p_{n+2} - (x + y)p_{n+1} + xyp_n > 0.
$$

That is

$$
p_{n+2} - xp_{n+1} > y(p_{n+1} - xp_n), n \geq 0.
$$

By $p_1 - xp_0 = 0$ and using the mathematical induction, we see that for $n > 0$ there is $p_{n+1} - xp_n > 0$. Combing with $p_2 > 0$ and using the mathematical induction, we see that for $n \geq 2$ there is $p_n > 0$.

Then we consider the case where $n < 0$ and let $q_m = p_{-m}$. Note that

$$\frac{1}{x} + \frac{1}{y} + \frac{1}{z} = \frac{xy + yz + zx}{xyz} > \frac{ab + bc + ca}{abc} = \frac{1}{a} + \frac{1}{b} + \frac{1}{c},$$

$$\frac{1}{xy} + \frac{1}{yz} + \frac{1}{zx} = \frac{x + y + z}{xyz} = \frac{a + b + c}{abc} = \frac{1}{ab} + \frac{1}{bc} + \frac{1}{ca},$$

$$\frac{1}{xyz} = \frac{1}{abc},$$

and then we have

$$q_{m+3} - \left(\frac{1}{a} + \frac{1}{b} + \frac{1}{c} \right) q_{m+2} + \left(\frac{1}{ab} + \frac{1}{bc} + \frac{1}{ac} \right) q_{m+1} - \frac{1}{abc} q_m$$

$$= \left(\frac{1}{a} + \frac{1}{b} + \frac{1}{c} - \frac{1}{x} - \frac{1}{y} - \frac{1}{z} \right) \left[x^{-(m+2)} + y^{-(m+2)} + z^{-(m+2)} \right] < 0.$$

And since $q_0 = q_{-1} = 0, q_1 = \frac{1}{a} + \frac{1}{b} + \frac{1}{c} - \frac{1}{x} - \frac{1}{y} - \frac{1}{z} < 0$, then by using a reasoning method similar to that for $n > 0$, we prove in turn by mathematical induction that

$$q_{m+2} - \left(\frac{1}{a} + \frac{1}{b} \right) q_{m+1} + \frac{1}{ab} q_m < 0, \ m \geq -1,$$

$$q_{m+1} - \frac{1}{a} q_m < 0, m \geq 0,$$

and for $m \geq 1$, there is $q_m < 0$.

In summary, the set of required ns required consists of the whole non-negative integers. $\square$

6 Suppose n, k are positive integers and $2 \leq n < 2^k$. Prove that: there exists a subset of $\{0, 1, 2, \ldots, n\}$, say A, that satisfies: for any $x, y \in A$, the combinatorial number C_y^x is even and

$$|A| \geq \frac{C_k^{\left[\frac{k}{2} \right]}}{2^k} \cdot (n + 1).$$

(Contributed by Qu Zhenhua)

Solution 1 For positive integer m, suppose its binary representation is

$$m = 2^{\alpha_1} + 2^{\alpha_2} + \cdots + 2^{\alpha_i}, \ \alpha_1 > \alpha_2 > \cdots > \alpha_t \geq 0.$$

Let $P_m = \{\alpha_1, \alpha_2, \ldots, \alpha_t\}$ and assume $P_0 = \emptyset$. By Lucas' theorem, C_y^x is even $\Leftrightarrow P_x \not\subset P_y$.

Consider set family $\mathcal{F} = \{P_0, P_1, \ldots, P_n\}$. As $n < 2^k$, then $P_i \subseteq \{0, 1, \ldots, k-1\}$, $0 \le i \le n$. The problem is equivalent to proving that there exists $\mathcal{A} \subset \mathcal{F}$, such that the sets in $\mathcal{A}$ do not contain each other, and

$$|\mathcal{A}| \ge \frac{C_k^{\left[\frac{k}{2}\right]}}{2^k} \cdot (n+1).$$

If $n + 1 = 2^k$, $\mathcal{F}$ consists of all the subsets of $\{0, 1, \cdots, k-1\}$. Then all the $\left[\dfrac{k}{2}\right]$-element subsets, which do not contain each other, can be taken to form set family $\mathcal{A}$. At this point, $|\mathcal{A}| = \dfrac{C_k^{\left[\frac{k}{2}\right]}}{2^k} \cdot (n+1)$.

In the following, we assume that $n + 1 < 2^k$.

Write $n + 1$ as a binary representation

$$n + 1 = 2^{k_1} + 2^{k_2} + \cdots + 2^{k_s}, \quad k > k_1 > k_2 > \cdots > k_s \ge 0.$$

And let $r_1 > r_2 > \cdots > r_s$ be a sequence of integers. We construct set famlity A as follows:

First, we specify some notations. For positive integer x, denote $\langle x \rangle$ as the x-element set $\{0, 1, \cdots, x-1\}$. $S(X, r)$ is the set consisting of all r-element subsets of set X. Assume that $S(X, r)$ is an empty set when $r > |X|$ or $r < 0$. For $2 \le i \le s$, we define set family

$$\mathcal{A}_i = \{X \cup \{k_1, \cdots, k_{i-1}\} : X \in S(\langle k_i \rangle, r_i)\},$$

and $\mathcal{A}_1 = S(\langle k_1 \rangle, r_1)$ for $i = 1$. Take

$$\mathcal{A} = \mathcal{A}_1 \cup \mathcal{A}_2 \cup \cdots \cup \mathcal{A}_s.$$

We show that the sets in $\mathcal{A}$ do not contain each other. Since the sets in $S(\langle k_1 \rangle, r_1)$ do not contain each other, then the sets in $\mathcal{A}_i$ also do not contain each other. For $1 \le i < j \le s$, $X \in \mathcal{A}_i, Y \in \mathcal{A}_j$. Since

$$|Y| = r_j + (j - 1) \le r_i + (i - 1) = |X|,$$

The only possible containing relation is $Y \subset X$. But $k_i \in Y, k_i \notin X$, so $Y \not\subset X$. Therefore, the sets in $\mathcal{A}$ do not contain each other, and then

$$|\mathcal{A}| = C_{k_1}^{r_1} + C_{k_2}^{r_2} + \cdots + C_{k_s}^{r_s}.$$

In the following we will show that we can take $r_1 > r_2 > \cdots > r_s$ such that

$$\frac{|\mathcal{A}|}{n+1} = \frac{C_{k_1}^{r_1} + C_{k_2}^{r_2} + \cdots + C_{k_s}^{r_s}}{2^{k_1} + 2^{k_2} + \cdots + 2^{k_s}} \ge \frac{C_k^{\left[\frac{k}{2}\right]}}{2^k}.$$

Let $r_1 = \left\lceil \dfrac{k_1}{2} \right\rceil$ (The smallest integer not less than $\dfrac{k_1}{2}$). For $i = 2, 3, \ldots, s$, let

$$r_i = \min \left\{ r_{i-1} - 1, \left\lceil \frac{k_i}{2} \right\rceil \right\}.$$

For $r_1 > r_2 > \cdots > r_s$, if $r_i = \left\lceil \dfrac{k_i}{2} \right\rceil$, then i is said to be a "good number". Suppose $1 = i_1 < \cdots < i_t$ are the whole good numbers. Let $i_{t+1} = s + 1$. It is sufficient to prove that for $j = 1, 2, \ldots, s$, there is

$$\frac{C_{k_{i_j}}^{r_{i_j}} + C_{k_{i_j+1}}^{r_{i_j+1}} + \cdots + C_{k_{i_{j+1}-1}}^{r_{i_{j+1}-1}}}{2^{k_{i_j}} + 2^{k_{i_j+1}} + \cdots + 2^{k_{i_{j+1}-1}}} \geq \frac{C_k^{\left[\frac{k}{2}\right]}}{2^k}.$$

Note that $r_{i_j}, r_{i_j+1}, \ldots, r_{i_{j+1}-1}$ is $i_{j+1} - i_j$ consecutive numbers. To simplify the notation, we use $r, r-1, \ldots, r-m, m \geq 0$ (where there may be negative numbers) instead. Rewrite $k_{i_j}, \ldots, k_{i_{j+1}-1}$ as

$$K_0 > K_1 > \cdots > K_m,$$

where $r = \left\lceil \dfrac{K_0}{2} \right\rceil$, $K_0 < k, K_m \geq 0$. We need to prove

$$\frac{C_{K_0}^r + C_{K_1}^{r-1} + \cdots + C_{K_m}^{r-m}}{2^{K_0} + 2^{K_1} + \cdots + 2^{K_m}} \geq \frac{C_k^{\left[\frac{k}{2}\right]}}{2^k}. \qquad \textcircled{1}$$

For the left side of Formula $\textcircled{1}$, denote the numerator as A and the denominator as B. We carry out the proof by induction on m. When $m = 0$,

$$\frac{A}{B} = \frac{C_{K_0}^{\left\lceil \frac{K_0}{2} \right\rceil}}{2^{K_0}} \geq \frac{C_k^{\left[\frac{K}{2}\right]}}{2^k},$$

and here we use the fact that $\dfrac{C_k^{\left[\frac{K}{2}\right]}}{2^k}$ is monotonic decreasing with respect to k.

Next we assume $m > 0$ and the conclusion holds for all cases less than m. Suppose $0 \leq l \leq m$ such that $K_l, \ldots, K_m$ are consecutive integers, but $K_{l-1} \geq K_l + 2$ (or $l = 0$). Take

$$A_1 = C_{K_0}^r + C_{K_1}^{r-1} + \cdots + C_{K_{-1}}^{r-m+1} = A - C_{K_m}^{r-m},$$

$$B_1 = 2^{K_0} + 2^{K_1} + \cdots + 2^{K_{m-1}} = B - 2^{K_m},$$

$$A_2 = \mathrm{C}_{K_0}^r + \cdots + \mathrm{C}_{K_{t-1}}^{r-l+1} + \mathrm{C}_{K_i+1}^{r-l},$$

$$B_2 = 2^{K_0} + \cdots + 2^{K_{t-1}} + 2^{K_t+1} = B + 2^{K_m}.$$

By using Pascal's identity repeatedly, we know

$$A_1 + A_2 \le 2A \Leftrightarrow \mathrm{C}_{K_l+1}^{r-l} \le \mathrm{C}_{K_l}^{r-l} + \cdots + \mathrm{C}_{K_{m-1}}^{r-m+1} + 2\mathrm{C}_{K_m}^{r-m}$$

$$\Leftrightarrow \mathrm{C}_{K_m-1}^{r-m-1} \le \mathrm{C}_{K_m}^{r-m}$$

$$\Leftrightarrow 0 \le \mathrm{C}_{K_m-1}^{r-m}.$$

As $B_1 + B_2 = 2B$, then

$$\frac{A}{B} \ge \frac{A_1 + A_2}{B_1 + B_2} \ge \min\left\{\frac{A_1}{B_1}, \frac{A_2}{B_2}\right\}.$$

If $\dfrac{A}{B} \ge \dfrac{A_1}{B_1}$, then by the induction hypothesis, we know that $\dfrac{A_1}{B_1} \ge \dfrac{\mathrm{C}_k^{\left[\frac{k}{2}\right]}}{2^k}$, and the conclusion holds.

If $\dfrac{A}{B} \ge \dfrac{A_2}{B_2}$, then we consider $\dfrac{A_2}{B_2}$, which is equivalent to replacing $K_l, \cdots, K_m$ with $K_l + 1$. If $K_l, \ldots, K_m$ contain at least two numbers, then $\dfrac{A_2}{B_2} \ge \dfrac{\mathrm{C}_k^{\left[\frac{k}{2}\right]}}{2^k}$ can still be obtained by the induction hypothesis and the conclusion holds. If $l = m$, the above discussion can be repeated on $\dfrac{A_2}{B_2}$, and the induction hypothesis can be used after some adjustments, and the conclusion holds.

Solution 2 The definition of $P_m = \{\alpha_1, \alpha_2, \ldots, \alpha_t\}$ is the same as in Solution 1. The problem is equivalent to proving that for set family $\mathcal{F} = \{P_0, P_1, \cdots, P_n\}$, there exists $\mathcal{A} \subset \mathcal{F}$ satisfying that the sets in $\mathcal{A}$ do not contain each other, and

$$|\mathcal{A}| \ge \frac{\mathrm{C}_k^{\left[\frac{k}{2}\right]}}{2^k} \cdot (n+1).$$

Since $\dfrac{\mathrm{C}_k^{\left[\frac{k}{2}\right]}}{2^k}$ is monotonic decreasing with respect to k, we only consider the smallest k that satisfies $2^k > n$, i.e.,

$$2^{k-1} \le n < 2^k.$$

We will prove a stronger proposition.

Proposition Let n, k be positive integers, $2^{k-1} \leq n < 2^k$, and t be an integer with $0 \leq t \leq \left[\dfrac{k}{2}\right]$. Then there exists set family $\mathcal{A} \subset \mathcal{F} = \{P_0, P_1, \ldots, P_n\}$ satisfying

(i) the sets in $\mathcal{A}$ do not contain each other;

(ii) the number of elements of each set in $\mathcal{A}$ is not greater than t;

(iii) $|\mathcal{A}| \geq \dfrac{C_k^{\left[\frac{k}{2}\right]}}{2^k} \cdot (n+1)$.

The original proposition can be derived when $t = \left[\dfrac{k}{2}\right]$. We will prove the proposition by induction on k.

When $k = 1$, $n = 1, t = 0$. It is sufficient that we simply take $\mathcal{A} = \{P_0 = \varnothing\}$.

In the following, we assume $k \geq 2$ and that the conclusion holds for each smaller k. Let $n = 2^{k-1} + n'$, then $0 \leq n' < 2^{k-1}$. If $t = 0$, we only need to take $\mathcal{A} = \{P_0 = \varnothing\}$. We consider the case $t > 0$. Since $P_0, P_1, \ldots, P_{2^{k-1}-1}$ is exactly the 2^{k-1} subsets of $\{0, 1, \ldots, k-2\}$, by $t \leq \left[\dfrac{k}{2}\right] \leq k - 1$, we take all t-element subsets $A_1, A_2, \ldots, A_p$ among them, where $p = C_{k-1}^t$. These sets do not contain each other. If n', namely, $n = 2^{k-1}$, take

$$\mathcal{A} = \{A_1, A_2, \ldots, A_p, P_{2^{k-1}} = \{k-1\}\},$$

and it is easy to find that $\mathcal{A}$ satisfies Conditions (i) and (ii). And there is

$$|\mathcal{A}| = C_{k-1}^t + 1 \geq \frac{1}{2} C_k^t + \frac{C_k^t}{2^k} = \frac{C_k^t}{2^k}(2^{k-1} + 1) = \frac{C_k^t}{2^k}(n+1).$$

Two simple inequalities, $C_{k-1}^t \geq \dfrac{1}{2} C_k^t$ and $1 \geq \dfrac{C_k^t}{2^k}$, are used here, with the former being equivalent to $2t \leq k$.

If $n' > 0$, we may let $2^{r-1} \leq n' < 2^r$, where $0 < r \leq k - 1$. Note that $P_{2^{k-1}}, P_{2^{k-1}+1}, \ldots, P_{2^{k-1}+n'}$ is just $P_0 \cup \{k-1\}, P_1 \cup \{k-1\}, \ldots, P_{n'} \cup \{k-1\}$. The discussion is divided into the following two cases.

Case 1: $t - 1 \leq \left[\dfrac{r}{2}\right]$. By induction hypothesis, there exists $\mathcal{B} \subset \{P_0, P_1, \ldots, P_{n'}\}$, where the sets in $\mathcal{B}$ do not mutually contain each other. The number of elements of each set in $\mathcal{B}$ is not greater than $t - 1$, and

$|\mathcal{B}| \geq \dfrac{\mathrm{C}_r^{t-1}}{2^r}(n+1)$. Take

$$\mathcal{A} = \{A_1, A_2, \ldots, A_p\} \cup \{X \cup \{k-1\} \mid X \in \mathcal{B}\},$$

and it is clear that $\mathcal{A}$ satisfies conditions (i) and (ii). Since

$$\frac{\mathrm{C}_r^{t-1}}{2^r} \geq \frac{\mathrm{C}_{r+1}^{t-1}}{2^{r+1}} \geq \cdots \geq \frac{\mathrm{C}_{k-1}^{t-1}}{2^{k-1}},$$

then

$$|\mathcal{A}| \geq \mathrm{C}_{k-1}^t + \frac{\mathrm{C}_r^{t-1}}{2^r}(n'+1) \geq \mathrm{C}_{k-1}^t + \frac{\mathrm{C}_{k-1}^{t-1}}{2^{k-1}}(n'+1).$$

Since $t \leq \dfrac{k}{2}$, then $t-1 < \dfrac{k-1}{2}$, and thus $\mathrm{C}_{k-1}^t \geq \mathrm{C}_{k-1}^{t-1}$. And as $1 \geq \dfrac{n'+1}{2^{k-1}}$, by Chebysheev inequality we have

$$\mathrm{C}_{k-1}^t + \frac{\mathrm{C}_{k-1}^{t-1}}{2^{k-1}}(n'+1) \geq \frac{1}{2}(\mathrm{C}_{k-1}^t + \mathrm{C}_{k-1}^{t-1})\left(1 + \frac{n'+1}{2^{k-1}}\right) = \frac{\mathrm{C}_k^t}{2^k}(n+1),$$

and hence $|\mathcal{A}| \geq \dfrac{\mathrm{C}_k^t}{2^k} \cdot (n+1)$.

Case 2　$t - 1 > \left[\dfrac{r}{2}\right]$. By induction hypothesis, there exists $\mathcal{B} \subset \{P_0, P_1, \ldots, P_{n'}\}$, where the sets in $\mathcal{B}$ do not mutually contain each other. The number of elements of each set in $\mathcal{B}$ is not greater than $\left[\dfrac{r}{2}\right]$, and

$|\mathcal{B}| \geq \dfrac{\mathrm{C}_r^{\left[\frac{r}{2}\right]}}{2^r}(n'+1)$. Take

$$\mathcal{A} = \{A_1, A_2, \ldots, A_p\} \cup \{X \cup \{k-1\} \mid X \in \mathcal{B}\},$$

and it is clear that $\mathcal{A}$ satisfies conditions (i) and (ii). Since

$$\frac{\mathrm{C}_r^{\left[\frac{r}{2}\right]}}{2^r} \geq \frac{\mathrm{C}_r^{\left[\frac{k-1}{2}\right]}}{2^{k-1}} \geq \frac{\mathrm{C}_{k-1}^{t-1}}{2^{k-1}},$$

then

$$|\mathcal{A}| \geq \mathrm{C}_{k-1}^t + \frac{\mathrm{C}_r^{\left[\frac{r}{2}\right]}}{2^r}(n'+1) \geq \mathrm{C}_{k-1}^t + \frac{\mathrm{C}_{k-1}^{t-1}}{2^{k-1}}(n'+1).$$

By the same method as in Case 1, we obtain $|\mathcal{A}| \geq \dfrac{\mathrm{C}_k^t}{2^k} \cdot (n+1)$.

Therefore, we have proved the stronger proposition.　□

China Girls' Mathematical Olympiad

2018 (Chengdu, Sichuan)

The 17th Chinese Girls' Mathematical Olympiad (CGMO) was held between August 11th and 16th, 2018 at Shude Secondary School, Chengdu, China. 148 female students from 37 teams from various provinces, municipalities, and autonomous regions across the mainland China, Hong Kong, Macau, Russia, the Philippines, and Singapore participated in the competition. After two rounds of contests (4 hours and 4 questions each), 36 students topped by Yi Ruizhe won the gold medals (first prize), 55 students topped by Liu Zhixuan won the silver medals (second prize), and 56 students topped by Zhong Weini won the bronze medals (third prize). In addition, the top 12 contestants including Yi Ruizhe are invited to participate in the 2018 National Mathematics Winter Camp (CMO).

Professor Chen Min, who is also vice president of the Academy of Mathematics and Systems Science, Chinese Academy of Sciences, vice chairman of the Chinese Mathematical Society and director of the Olympic Committee of the Chinese Mathematical Society, serves as the director of the organizing committee.

Director of the main examination committee: Li Shenghong.

Members of the main examination committee (in alphabetical order):

Ai Yinghua (Tsinghua University);

Fu Yunhao (Guangdong Second Normal University);

Ji Chungang (Nanjing Normal University);

Leng Fusheng (Chinese Academy of Sciences);
Li Shenghong (Zhejiang University);
Liang Yingde (University of Macau);
Lin Qiushi (Peking University);
Wang Bin (Chinese Academy of Sciences);
Wang Xinmao (University of Science and Technology of China).

First Day
(8:00 – 12:00; August 13, 2018)

1 Suppose real numbers $a \leq 1$, $x_0 = 0$, $x_{n+1} = 1 - ae^{x_n} (n \geq 0)$, where e is the base of the natural logarithm. Prove that for any positive integer n, $x_n \geq 0$.

Solution When $a \leq 0$, we have $x_n \geq 1$ for any $n \geq 1$. When $0 < a \leq 1$, note that $x_1 = 1 - a$, and $f(x) = 1 - ae^x$ is a monotonic decreasing function. Suppose $x_n \in [0, 1 - a]$, then

$$x_{n+1} = f(x_n) \in [f(1 - a), f(0)].$$
$$f(0) = 1 - a,$$
$$f(1 - a) = 1 - \frac{a}{e^{a-1}} \geq 1 - \frac{a}{1 + (a - 1)} = 0.$$

Therefore, $x_{n+1} \in [0, 1 - a]$. Consequently, for any positive integer n, $x_n \geq 0$. □

2 As shown in Fig. 2.1, suppose points D, E are on sides AB, AC of $\triangle ABC$, respectively, and $DE//BC$. O_1, O_2 are the circumcenters of $\triangle ABE$, $\triangle ACD$, respectively. Line O_1O_2 intersects AB at point P and intersects AC at point Q. O is the circumcenter of $\triangle APQ$, and M is the intersection of lines AO and BC.
Prove that M is the midpoint of BC.

Solution 1 As shown in Fig. 2.2, let F, G, J, K be the midpoints of AD, AE, AB, AC, respectively. Connecting O_1J, O_1G, O_2F, O_2K, then we have $AJ = \frac{1}{2}AB$, $AF = \frac{1}{2}AD$, and thus $FJ = \frac{1}{2}BD$. Similarly, we have $GK = $

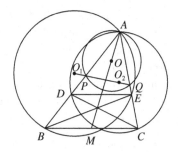

Fig. 2.1

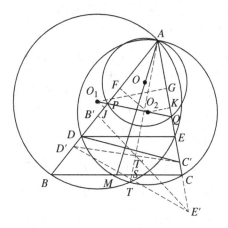

Fig. 2.2

$\frac{1}{2}CE$. In addition, as $FJ = O_1O_2 \cos \angle APQ$, $GK = O_1O_2 \cos \angle AQP$, thus

$$\frac{\cos \angle APQ}{\cos \angle AQP} = \frac{FJ}{GK} = \frac{BD}{CE} = \frac{AB}{AC}.$$

Since O is the circumcenter of $\triangle APQ$,

$$\angle OAQ = \frac{\pi}{2} - \angle APQ, \angle OAP = \frac{\pi}{2} - \angle AQP,$$

then we have $\dfrac{BM}{CM} = \dfrac{AB \sin \angle OAP}{AC \sin \angle OAQ} = \dfrac{AB \cos \angle AQP}{AC \cos \angle APQ} = 1$. Therefore, M is the midpoint of BC.

Solution 2 As shown in Fig. 2.2, since $DE//BC$, then $AC \cdot AD = AB \cdot AE$. We perform an inversion with A as the inversion center and $\sqrt{AC \cdot AD}$ as the inversion radius. B', C', D', E' are the images of B, C, D, E under the inversion, respectively, and it is easy to find $AB' = AE$, $AC' = AD$, $AD' = AC$, $AE' = AB$. Let T be the second intersection of $\triangle ABE$ and the circumscribed circle of $\triangle ACD$. Then the image T' of T under the inversion is the intersection of $B'E'$, $C'D'$.

Since A, T, T' are collinear, then $O_1O_2 \perp AT'$. From the familiar conclusion, we know that AO and AT' are isotomic conjugate with respect to $\angle BAC$. Extend AT' to intersect $D'E'$ at S. By Ceva theorem, there is $\dfrac{AB'}{B'D'} \cdot \dfrac{D'S}{SE'} \cdot \dfrac{E'C'}{C'A} = 1$, and thus $D'S = SE'$, i.e., S is the midpoint of $D'E'$. Since AO and AT' are symmetric about the angle bisector of $\angle BAC$, $\triangle ABC$ and $\triangle AE'D'$ are symmetric with respect to the angle bisector of $\angle BAC$, so M and S are symmetric regarding the angle bisector of $\angle BAC$. Therefore, M is the midpoint of BC.

Solution 3 As shown in Fig. 2.3, suppose another intersection of circles O_1 and O_2 that are different from point A is T. Join AT, BT, CT, DT, ET.

Since A, C, T, D are concyclic, $\angle BDT = \angle ECT$. Similarly, $\angle CET = \angle DBT$, and then $\triangle BDT \sim \triangle ECT$. Therefore, $\dfrac{BT}{ET} = \dfrac{BD}{EC} = \dfrac{AB}{AC}$. Since A, B, T, E are concyclic, then $\dfrac{BT}{\sin \angle BAT} = \dfrac{ET}{\sin \angle EAT}$.

Obviously, $O_1O_2 \perp AT$, and thus $\angle EAT + \angle AQP = \dfrac{\pi}{2}$. O is the circumcenter of $\triangle APQ$, so $\angle OAP + \angle AQP = \dfrac{\pi}{2}$, and hence $\angle OAP = \angle EAT$. Similarly, there is $\angle OAQ = \angle BAT$. Therefore, we have $\dfrac{BM}{CM} = \dfrac{AB \sin \angle OAP}{AC \sin \angle OAQ} = \dfrac{BT \sin \angle EAT}{ET \sin \angle BAT} = 1$. Consequently, M is the midpoint of BC. □

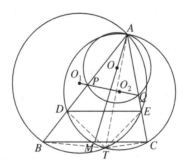

Fig. 2.3

3 Suppose sequence of real numbers $\{x_n\}_{n=1}^{\infty}$ satisfies $x_1 = 1$. Prove that: for any integer $n \geq 2$, there is

$$\sum_{i|n} \sum_{j|n} \frac{x_i x_j}{[i,j]} \geq \prod_{\text{prime } p|n} \left(1 - \frac{1}{p}\right),$$

where $\sum_{i|n}$ denotes the sum of all the positive factors i of n, $[i,j]$ indicates the least common multiple of i, j, and $\prod_{\text{prime } p|n}$ is the product of all the prime factors p of n.

Solution We should notice that the number of integers from 1 to n that are divisible by $[i,j]$ for factors i, j of n is $\dfrac{n}{[i,j]}$, and there is

$$\frac{n}{[i,j]} = \sum_{[i,j]|k,\, k \leq n} 1.$$

From this it follows that

$$\sum_{i|n,j|n} x_i x_j \frac{n}{[i,j]} = \sum_{i|n,j|n} x_i x_j \sum_{[i,j]|k,k\leq n} 1 = \sum_{i|n,j|n} x_i x_j \sum_{i|k,j|k,k\leq n} 1$$

$$= \sum_{k\leq n} \sum_{i|(k,n)} \sum_{j|(k,n)} x_i x_j = \sum_{k\leq n} \left(\sum_{i|(k,n)} x_i \right)^2$$

$$\geq \sum_{k\leq n,(k,n)=1} \left(\sum_{i|(k,n)} x_i \right)^2 = \sum_{k\leq n,(k,n)=1} x_1^2 = \phi(n),$$

where $\phi(n)$ is Euler function, i.e., the numbers in $1, \ldots, n$ that are coprime to n. It is familiar that

$$\phi(n) = n \prod_{\text{prime } p|n} \left(1 - \frac{1}{p}\right).$$

Therefore, we have completed the proof. □

4 n students with different names send greeting cards to each other. Each person prepares $n - 1$ envelopes each with the names and addresses of the other $n - 1$ students, respectively, and also prepares at least one greeting card signed with his/her own name. Each day, one student will send a greeting card as follows: choose a greeting card he/she owns, including the one he/she received before, put it into an unused envelope so that the signature of the card is different from the name of the recipient, and then send the card, which will be received on the same day.

Prove that: when all the students are unable to send cards as described above,

(1) each has at least one card in hand;
(2) if at this time there exist k students $P_1, P_2, \ldots, P_k$ such that P_k has never sent a card to P_1 and P_i has never sent a card to P_{i+1} $(i = 1, \ldots, k-1)$, then these k students initially prepared the same number of cards.

Solution Let the n students be $P_1, P_2, \ldots, P_n$, the initial numbers of greeting cards in their hands be $p_1, p_2, \ldots, p_n$, and the final numbers of cards they own be $q_1, q_2, \ldots, q_n$, respectively.

First, note that if P_i has never sent cards to P_j at the final moment, then all the greeting cards in P_i's hand are with the signature of P_j. Moreover, if $q_i > 0$, then P_j is the only person P_i did not send cards to him/her.

Given a subset V_0 of set $V = \{P_1, \ldots, P_n\}$, at the end of any day, let the total number of unsent envelopes from V_0 to $V \backslash V_0$ be r_{V_0}, the total number of unsent envelopes from $V \backslash V_0$ to V_0 be s_{V_0}, and the total number of greeting cards in V_0 be m_{V_0} (all three are functions of the number of days). Note that for each passed day, one of the following three things must occur:

(i) r_{V_0} and m_{V_0} decrease by 1 at the same time, and s_{V_0} remains unchanged.
(ii) s_{V_0} decreases by 1, m_{V_0} increases by 1, and r_{V_0} remains unchanged.
(iii) $r_{V_0}, s_{V_0}, m_{V_0}$ remain unchanged.

Therefore, $m_{V_0} - r_{V_0} + s_{V_0}$ is invariant.

Denote the set of all the greeting cards in V_0 at the initial moment as pV_0, and that at the final moment as qV_0. Then $|p(V_0)| = \sum_{P_i \in V_0} p_i$, $|q(V_0)| = \sum_{P_i \in V_0} q_i$. Comparing the numbers of cards between the initial and final moments shows that

$$|p(V_0)| = |q(V_0)| - r_{V_0} + s_{V_0}. \qquad \text{①}$$

Lemma *There does not exist $V_0 \subseteq V$ such that $q(V_0) \subset p(V_0)$.*

Proof of lemma We will use proof by contradiction. Assume there exists such $V_0 \subseteq V$. Since $|p(V_0)| < |q(V_0)|$, by ① we have $s_{V_0} > r_{V_0} \geq 0$. Then there exists P_{i_1} in V_0 and P_{j_1} in $V \backslash V_0$ such that P_{j_1} never sends a card to P_{i_1}. So P_{j_1} has only the greeting cards signed by P_{i_1} in his hand. Let $V_1 = V_0 \cup \{P_{j_1}\}$. Then $q(V_1) \subseteq p(V_0) \subset p(V_1)$, and hence there exists some P_{i_2} in V_1, and some P_{j_2} in $V \backslash V_0$ such that P_{j_2} never sends a letter to P_{i_2}. This process can be repeated infinitely, leading to a contradiction.

We go back to the original question.

(1) We will use proof by contradiction. Suppose there exists P_i such that $q_i = 0$. Let $V_0 = \{P_i\}$, and then $q(V_0) \subset p(V_0)$. This is in contradiction to the lemma. Therefore, there is no q_i equal to 0.

(2) Let $V_0 = \{P_1, \ldots, P_k\}$. Notice that $q(\{P_i\}) \subseteq p(\{P_{i+1}\})$, so we have $q_1 \leq p_2, \ldots, q_k \leq p_1$, and also $q(V_0) \subseteq p(V_0)$. From the lemma, it follows that $q(V_0) = p(V_0)$, so all the above inequalities take equal signs. This means that each q_i is greater than 0, and thus P_i has only one unsent envelope (with the name of P_{i+1}). Therefore, $r_{\{P_i\}} = 1$, and $r_{V_0} = 0$. From ① we learn that $s_{V_0} = 0$, so for every P_i there is $s_{\{P_i\}} = 1$. Lastly, from ① it follows that $p_i = q_i$ $(i = 1, 2, \ldots, k)$. Combining $q_1 = p_2, \ldots, q_k = p_1$, we find $p_1 = p_2 = \cdots = p_k$. The proof is completed. $\square$

Second Day
(8:00 – 12:00; August 14, 2018)

5 Suppose complex number $z = (w+2)^3(w-3)^2$. Find the maximum value of the modulus of z when w is taken over all the complex numbers of modulus 1.

Solution If α is a real number and w is a complex number of modulus 1, let $t = w + \bar{w} = 2\operatorname{Re} w$. Then

$$|w - \alpha|^2 = (w - \alpha)(\bar{w} - \bar{\alpha}) = |w|^2 - \alpha(w + \bar{w}) + \alpha^2 = 1 - \alpha t + \alpha^2.$$

By the inequality of arithmetic and geometric means, it follows that

$$|z|^2 = |(w+2)^3(w-3)^2|^2 = (|w+2|^2)^3(|w-3|^2)^2$$

$$= (5 + 2t)^3(10 - 3t)^2 \leq \left[\frac{3(5 + 2t) + 2(10 - 3t)}{3 + 2}\right]^{3+2} = 7^5.$$

The equal sign holds if and only if $5 + 2t = 10 - 3t$, namely, $t = 1$. Accordingly, $w = \dfrac{1 \pm \sqrt{3}\mathrm{i}}{2}$, and the maximum value of $|z|$ is $7^{\frac{5}{2}}$. $\square$

6 Given positive integer k. If sequence $I_1 \supseteq I_2 \supseteq \cdots \supseteq I_k$ of subsets of set $\mathbb{Z}$ of integers satisfies $168 \in I_i$ for $i = 1, 2, \ldots, k$, and for any

$x, y \in I_i$ (where x and y can be equal) there is $x - y \in I_i$, then $(I_1, I_2, \ldots, I_k)$ is said to be a k-chain in $\mathbb{Z}$. Question: How many k-chains are there in $\mathbb{Z}$? Please explain the justification.

Solution　Let I be a non-empty subset of $\mathbb{Z}$ and there is $x - y \in I$ for any $x, y \in I$. I is said to be a "good set", if either $I = \{0\}$ or $I = \{a \mid a \in \mathbb{Z}$ and $b \mid a\}$, where b is a positive integer. In addition, if I is a "good set", let b be the smallest positive integer in I. Then we have $b \mid a$ for any $a \in I$. Otherwise, there is $b \nmid a$. By the division algorithm, there exist unique integers q, r such that $a = bq + r$, where $1 \leq r \leq b - 1$. Then $r = a - bq \in I$, contradicting the minimality of b. Therefore, $I \neq \{0\}$ is a "good set" if and only if there exists positive integer b such that $I = \{a \mid a \in \mathbb{Z}$ and $b \mid a\}$. At this point, denote $I = b\mathbb{Z}$.

We may regard $(I_1, I_2, \ldots, I_k)$ as a k-chain on $\mathbb{Z}$. Since $168 \in I_i$, then $I_i \neq 0$. Let $I_i = b_i\mathbb{Z}$, where b_i is a positive integer and $b_i \mid 168$. As $168 = 2^3 \cdot 3 \cdot 7$, then $b_i = 2^{x_i} \cdot 3^{y_i} \cdot 7^{z_i}$. Since $I_1 \supseteq I_2 \supseteq \cdots \supseteq I_k$, we get $b_i \mid b_{i+1}$, $i = 1, 2, \ldots, k - 1$. Therefore,

$$0 \leq x_1 \leq \cdots \leq x_k \leq 3; \quad 0 \leq y_1 \leq \cdots \leq y_k \leq 1; \quad 0 \leq z_1 \leq \cdots \leq z_k \leq 1.$$

If $1 \leq w_1 \leq w_2 \leq \cdots \leq w_k \leq a$, let $t_1 = w_1$, $t_2 = w_2 - w_1, \ldots, t_k = w_k - w_{k-1}$, $t_{k+1} = a - w_k$. Then

$$t_1 + t_2 + \cdots + t_{k+1} = a,$$

where $t_i \in \mathbb{N}$, $i = 1, 2, \ldots, k + 1$. Thus the number of non-negative integer solutions of this equation is $C_{k+a}^k = C_{k+a}^a$.

Therefore, the number of k-chains on $\mathbb{Z}$ is

$$C_{k+3}^3 C_{k+1}^1 C_{k+1}^1 = \frac{(k+1)^3(k+2)(k+3)}{6}. \qquad \square$$

7　A grid has 2018 rows and 4 columns. Color each cell of the grid with red or blue so that the number of red cells in each row of the grid is equal to the number of blue ones, and the number of red cells in each column is equal to the number of blue ones too. Denote M as the total number of coloring methods that satisfy the above conditions. Find the remainder of M divided by 2018.

Solution 1　We call the coloring method that satisfies the condition a "pattern". All the patterns form set $\mathbb{M}$.

We use (i, j) to denote the cell in row i and column j, and treat row i as a modulus of 2018, i.e., row 2018 can be regarded as row 0. For a given pattern A, we can do operation T to re-color it in this way: color (i, j) with the previous color in $(i - 1, j)$. After the operation, row 1 becomes row 2, row 2 becomes row 3, ..., row 2018 becomes row 1, while the column remains unchanged. The resulted pattern $T(A)$ also satisfies the conditions of the question, so it is also in $\mathbb{M}$.

Operation T is a one-to-one mapping from $\mathbb{M}$ to itself. Each element A of $\mathbb{M}$ returns to itself after 2018 operations T, i.e., $T^{(2018)}(A) = A$. We consider the period of A, i.e., the smallest positive integer $d = d(A)$ such that $T^{(d)}(A) = A$, and call the d-element set $R(A) = \{A, T(A), \ldots, T^{(d-1)}(A)\}$ the orbit of A. It is obvious that the orbits generated by other elements in $R(A)$ are also $R(A)$. For two elements A_1, A_2 in $\mathbb{M}$, if $R(A_1) \cap R(A_2) \neq \varnothing$, then $R(A_1) = R(A_2)$, i.e., any two orbits either coincide or do not intersect, and $\mathbb{M}$ can be regarded as a non-intersecting union of several orbits.

The period $d = d(A)$ must be a factor of 2018, i.e., one of $1, 2, 1009, 2018$. Since exactly half of the 2018 cells in each column are red, then $T^{(1009)}(A) \neq A$. Otherwise, the red cells can be paired, and this will lead to a contradiction to the parity. Hence the period of $d = d(A)$ is either 2 or 2018. If $d(A) = 2$, then all the odd rows of A are identical. And in the meantime, and all even rows are also identical and have the opposite coloring of the odd rows. The first row of A has $C_4^2 = 6$ coloring and the first row of A determines all other rows. Therefore, there are exactly 6 elements with period 2 in $\mathbb{M}$. The rest of the elements are all with period 2018.

Viewing $\mathbb{M}$ as a non-intersecting union of orbits, there are 3 orbits of length 2 and several orbits of length 2018. Therefore, the remainder of the number of elements $M = |\mathbb{M}|$ divided by 2018 is 6.

Solution 2 $2018 = 2 \times 1009$, let $p = 1009$ (a prime number).

The 4 cells in each row can be colored in $C_4^2 = 6$ ways. Let the number of rows in which two cells i, j are colored with red and the other two cells are colored with blue be x_{ij}, where $i < j$. Considering the number of red cells in each column, four equations about 6 variables can be obtained:

$$\begin{cases} x_{12} + x_{13} + x_{14} = p, \\ x_{12} + x_{23} + x_{24} = p, \\ x_{13} + x_{23} + x_{34} = p, \\ x_{14} + x_{24} + x_{34} = p. \end{cases}$$

Note that the sum of the 6 variables is $2p$, and simplification gives $x_{12} = x_{34}$, $x_{13} = x_{24}$, $x_{14} = x_{23}$. By using parameters a, b, c, all the non-negative solutions of the equations can be presented as:

$$x_{12} = x_{34} = a, \quad x_{13} = x_{24} = b, \quad x_{14} = x_{23} = c,$$

$$a + b + c = p, \quad a, b, c \in \mathbb{N}.$$

In the case of a fixed set of $a + b + c = p$, this corresponds to 6 types of rows with a, a, b, b, c, c, respectively, and the number of methods of arranging them in a column is:

$$\binom{2p}{a, a, b, b, c, c} = \binom{2p}{p} \binom{p}{a, b, c}^2.$$

We consider the total number of methods M:

$$M = \sum_{a+b+c=p} \binom{2p}{a, a, b, b, c, c} = \binom{2p}{p} \sum_{a+b+c=p} \binom{p}{a, b, c}^2.$$

Since

$$\binom{2p}{p} = 2 \times \frac{(2p - 1) \times \cdots \times (p + 1)}{(p - 1) \times \cdots \times 1} \bmod p = 2,$$

and $\binom{p}{a, b, c} = \dfrac{p!}{a!b!c!}$ is not divisible by p only at $(a, b, c) = (0, 0, p)$, $(0, p, 0)$, $(p, 0, 0)$ (in these cases, it is equal to 1). It is easy to find that M mod $p = 6$.

On the other hand, $\binom{2p}{p}$ is even, and M is also even. Therefore, M divided by $2p(= 2018)$, and the remainder is 6. □

8 As shown in Fig. 8.1, point I is the incenter of $\triangle ABC$, D and E are the tangent points of the inscribed circle of $\triangle ABC$ on sides AB and AC, respectively. Line BI intersects AC at point F, line CI intersects AB at G, line DE intersects BI at M, line DE intersects CI at N, line DE intersects FG at P, and line BC intersects IP at Q.
Prove that $BC = 2MN$ if and only if $IQ = 2IP$.

Solution 1 As shown in Fig. 8.2, suppose $\angle BAC = \alpha$, $\angle BIG = \angle IBC + \angle ICB = \dfrac{\pi - \alpha}{2}$, $\angle NDG = \angle ADE = \dfrac{\pi - \alpha}{2} = \angle NIB$. Therefore, N, D, I, B are concyclic. Consequently, $\angle INB = \angle IDB = \dfrac{\pi}{2}$ and

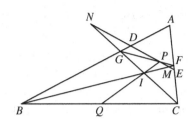

Fig. 8.1

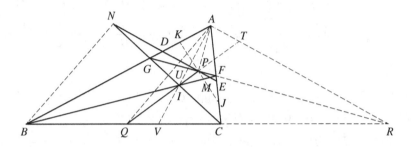

Fig. 8.2

$\angle DNI = \angle DBI = \angle IBC$. Thus

$$\triangle NIM \sim \triangle BIC, \quad \frac{NM}{BC} = \frac{NI}{BI} = \sin\frac{\alpha}{2}.$$

Suppose $BJ \perp AC$ at point J and $CK \perp AB$ at point K. Since CG is the angle bisector, $\frac{AG}{BG} = \frac{AC}{BC}$. Let $BC = a, AC = b, AB = c$, and then

$$AG = \frac{bc}{a+b}, \quad \frac{GI}{IC} = \frac{AG}{AC} = \frac{c}{a+b}, \quad \frac{GI}{GC} = \frac{c}{a+b+c}.$$

Similarly, there is $\frac{FI}{FB} = \frac{b}{a+b+c}$. It is easy to find $ID//CK$, and hence $\frac{DG}{KG} = \frac{GI}{GC}$. By the same reasoning, we can find $\frac{FJ}{EF} = \frac{FB}{FI}$.

Since B, K, C, J are concyclic, then $\triangle AJK \sim \triangle ABC, \frac{AK}{JA} = \frac{AC}{BA} = \frac{b}{c}$. Then applying Menelaus' theorem to $\triangle AGF$ and transversal DPE, we get $\frac{AD}{DG} \cdot \frac{GP}{PF} \cdot \frac{FE}{EA} = 1$. Since $AD = AE$, then $\frac{DG}{EF} = \frac{GP}{PF}$, and hence $\frac{AK}{JA} \cdot \frac{DG}{KG} \cdot \frac{FJ}{EF} = 1 = \frac{AK}{KG} \cdot \frac{GP}{PF} \cdot \frac{FJ}{JA}$. Applying the converse of Menelaus' theorem, we learn that K, P, J are collinear.

Suppose line FJ intersects BC at point R and connects AR. Line AI intersects FG at point U and BC at point V. By the properties of complete quadrangle, B, V, C, R are harmonic points, and there is $\dfrac{BV}{VC} = \dfrac{AB}{AC} = \dfrac{BR}{RC}$. Therefore, AR is exterior angle bisector of $\angle BAC$, and hence $AR \perp AI$. It is easy to learn that A, U, I, V are harmonic points and RA, FG, RI, BC are harmonic lines. Suppose line PQ intersects AR at point T. Then A, U, I, V are harmonic points. By Apollonius circle theorem, AI bisects PAQ.

Therefore, $\angle BAQ = \angle CAP$. And since $\angle AJK = \angle ABC$, then $\triangle AJP \sim \triangle ABQ$, $\dfrac{AP}{AQ} = \dfrac{AJ}{AB} = \cos\alpha$. As AI bisects PAQ, then $\dfrac{IP}{IQ} = \dfrac{AP}{AQ} = \cos\alpha$. In conclusion, we can get $\dfrac{MN}{BC} = \sin\dfrac{\alpha}{2} = \dfrac{1}{2} \Leftrightarrow \alpha = \dfrac{\pi}{3} \Leftrightarrow \dfrac{IP}{IQ} = \cos\alpha = \dfrac{1}{2}$. $\qquad\square$

Solution 2   As shown in Fig. 8.3, suppose $\angle BAC = \alpha$, $\angle BIG = \angle IBC + \angle ICB = \dfrac{\pi - \alpha}{2}$, $\angle NDG = \angle ADE = \dfrac{\pi - \alpha}{2} = \angle NIB$.

Then N, D, I, B are concyclic. Therefore, $\angle INB = \angle IDB = \dfrac{\pi}{2}$ and $\angle DNI = \angle DBI = \angle IBC$. Thus

$$\triangle NIM \sim \triangle BIC, \frac{NM}{BC} = \frac{NI}{BI} = \sin\frac{\alpha}{2}.$$

Suppose line FG intersects BC at point R and connects AR. Line AI intersects FG at U and BC at V. By the properties of complete quadrangle, B, V, C, R are harmonic points, and there is $\dfrac{BV}{VC} = \dfrac{AB}{AC} = \dfrac{BR}{RC}$. Therefore, AR is exterior angle bisector of $\angle BAC$, and hence $AR \perp AI$. It is easy to learn that A, U, I, V are harmonic points and RA, FG, RI, BC are harmonic lines. Suppose line PQ intersects AR at point T. Then T, P, I, Q are harmonic points. Therefore, we can let $\dfrac{IP}{IQ} = \dfrac{TP}{TQ} = x$, and it is easy to

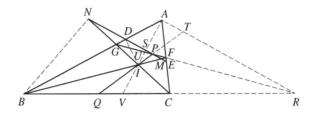

Fig. 8.3

find $\dfrac{IP}{TP} = \dfrac{1-x}{1+x}$. We assume AI intersects DE at point S. Since points D and E are symmetric about the line AI, then $DE \perp AI$, and hence $PS//AT$.

In conclusion, we find $\dfrac{MN}{BC} = \sin\dfrac{\alpha}{2} = \dfrac{1}{2} \Leftrightarrow \alpha = \dfrac{\pi}{3} \Leftrightarrow \dfrac{IP}{IQ} = \cos\alpha = \dfrac{1}{2}$. □

China Girls' Mathematical Olympiad

2019 (Wuhan, Hubei)

The 18th Chinese Girls' Mathematical Olympiad (CGMO) was held between August 11 and 14th, 2019 at Wuhan Foreign Languages School, Wuhan, China. 156 female students from 40 teams from various provinces, municipalities, and autonomous regions across the mainland China, Hong Kong, Macau, Russia, the Philippines, and Singapore participated in the competition. After two rounds of contests (4 hours and 4 questions each), 37 students topped by Yan Binwei won the gold medals (first prize), 59 students topped by Xie Xilai won the silver medals (second prize), and 60 students topped by Li Jinyan won the bronze medals (third prize). In addition, the top 14 contestants including Yan Binwei are qualified for participating in the 2019 National Mathematics Winter Camp (CMO).

Director of the main examination committee: Yu Hongbin (Suzhou University).

Members of the main examination committee (in alphabetical order):

Ai Yinghua (Tsinghua University);

Fu Yunhao (Southern University of Science and Technology);

Ji Chungang (Nanjing Normal University);

Leng Fusheng (Chinese Academy of Sciences);

Liang Yingde (University of Macau);

Lin Tianqi (East China Normal University);

Qu Zhenhua (East China Normal University);

Wang Xinmao (University of Science and Technology of China).
Wu Yuchi (East China Normal University);
Xiong Bin (East China Normal University).

First Day
(8:00 – 12:00; August 12, 2019)

1 As shown in Fig. 1.1, quadrilateral $ABCD$ is inscribed to $\odot O$. The tangents to $\odot O$ are made through points B, C and intersect at point L. Let M be the midpoint of BC. Construct a parallel line of AM through D and intersect the circumscribed circle of ADL at another point P. Suppose AP intersects $\odot O$ at another point E and DM at point K. Prove that $DK = EK$. (Contributed by Lin Tianqi)

Solution As shown in Fig. 1.2, since $\angle OBL = 90°$ and $BM \perp OL$ at point M, then

$$OM \cdot OL = OB^2 = OA^2 = OD^2.$$

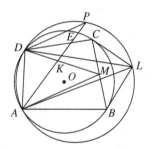

Fig. 1.1

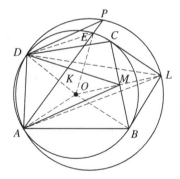

Fig. 1.2

Therefore, $\triangle OAM \sim \triangle OLA$, $\triangle ODM \sim \triangle OLD$, and then $\angle OAM = \angle OLA$, $\angle ODM = \angle OLD$. Combing with the fact that A, D, P, L are concyclic and $DP//AM$, it follows that

$$\angle ODM = \angle OLD = \angle ALD - \angle ALO$$

$$= \angle APD - \angle OAM$$

$$= \angle PAM - \angle OAM$$

$$= \angle PAO = \angle OEA.$$

Since $\angle ODE = \angle OED$, then $\angle KDE = \angle KED$, and hence $DK = EK$. $\qquad\square$

2 Please provide positive integers $a_1, a_2, \ldots, a_{18}$ such that $a_1 = 1$, $a_2 = 2$, $a_{18} = 2019$, and for any k $(3 \le k \le 18)$ there exist i, j $(1 \le i < j < k)$ such that $a_k = a_i + a_j$. (Contributeded by Wang Xinmao)

Solution Notice the Fibonacci sequence $F_0 = F_1 = 1, F_{n+2} = F_{n+1} + F_n$ satisfies $\dfrac{F_n}{F_{n+1}} \to \dfrac{\sqrt{5}-1}{2}$ when $n \to \infty$. We take $a_{17} = 1248$ to be the integer closest to $\dfrac{\sqrt{5}-1}{2}a_{18}$, $a_{16} = 2019 - 1248 = 771$, $a_{15} = 1248 - 771 = 477$, $a_{14} = 771 - 477 = 294$, $a_{13} = 477 - 294 = 183$, $a_{12} = 294 - 183 = 111$, $a_{11} = 183 - 111 = 72$, $a_{10} = 111 - 72 = 39$, $a_9 = 72 - 39 = 33$. Then take $a_{18} = 20$ as the integer closest to $\dfrac{\sqrt{5}-1}{2}a_9$, $a_7 = 33 - 20 = 13$, $a_6 = 20 - 13 = 7$, $a_5 = 13 - 7 = 6$. Since $a_1 = 1$, $a_2 = 2$, then $a_3 = 3$. Let $a_4 = a_2 + a_3 = 5$. From $a_5 = a_1 + a_4$ and $a_{10} = a_5 + a_9$, we know that the above constructed sequence $1, 2, 3, 5, 6, 7, 13, 20, 33, 39, 72, 111, 183, 294, 477, 771, 1248, 2019$ satisfies the given conditions. $\qquad\square$

3 The following operation can be performed on any sequence: each time you select three adjacent items in the sequence, denoted a, b, c, and replace them with b, c, a, while keeping the other items unchanged. Try to determine all the integers $n \ge 3$ such that sequence $1, 2, \ldots, n$ can be turned into $n, n-1, \ldots, 1$ by finite operations. (Contributed by Liang Yingde)

Solution Integer $n \ge 3$ is qualified if and only if $n \equiv 0$ or $1 \pmod 4$.

On one hand, for a sequence of m different numbers $\sigma_1, \sigma_2, \ldots, \sigma_n$ we define the inversion number of the sequence as

$$I(\sigma_1, \sigma_2, \ldots, \sigma_m) = |\ \{(j,k) : 1 \le j < k \le m \text{ and } \sigma_j > \sigma_k\}\ |\ .$$

If σ becomes τ after one time of the operation as described in the problem, it is easy to verify directly that

$$I(\tau) - I(\sigma) = I(b,c,a) - I(a,b,c) \in \{-2, 0, 2\}.$$

In particular, the parity of $I(\sigma)$ is unchanged after one operation. Thus, if we can change $1, 2, \ldots, n$ into $n, n-1, \ldots, 1$ after a series of the operation, then $I(1, 2, \ldots, n)$ and $I(n, n-1, \ldots, 1)$ must have the same parity, i.e., $n \equiv 0$ or $1 \pmod 4$.

On the other hand, we prove that when $n \equiv 0$ or $1 \pmod 4$, it is possible to change $1, 2, \ldots, n$ into $n, n-1, \ldots, 1$ after several times of the operation. We call this assertion $(\star)$. We will prove this assertion by induction on n. When $n = 4$, it is sufficient to operate in the following way:

$$1,2,3,4 \to 2,3,1,4 \to 2,1,4,3 \to 2,4,3,1 \to 4,3,2,1.$$

Assume that assertion $(\star)$ holds for $n < m$. In the following we consider the case $n = m$, where $m \equiv 0$ or $1 \pmod 4$, $m > 4$. Notice that operation $a, b, c \to b, c, a$ results in moving a from the left side of bc to its right side. First, move $m - 4$ to the right of $m - 3, m - 2, m - 1, m$ by two successive operations like this. Then move $m - 5$ to the right of $m - 3, m - 2, m - 1, m$ by two successive operations. Continuing doing this until 1 is moved to the right of $m - 3, m - 2, m - 1, m$, then we obtain sequence

$$m - 3, m - 2, m - 1, m, 1, 2, \ldots, m - 4.$$

Using the conclusion of the assertion $(\star)$ for $n = 4$, the first four terms $m - 3, m - 2, m - 1, m$ can be changed to the inversion order after several operations. For the remaining $m - 4$ terms, note that $m - 4 \equiv 0$ or $1 \pmod 4$. If $m - 4 = 1$, then no further operations are needed. If $m - 4 \ge 4$, by the induction hypothesis, $1, 2, \ldots, m - 4$ becomes the inversion order after several operations. Combing the above results, we prove that the assertion $(\star)$ holds for $n = m$. □

4 Given parallelogram $OABC$ in the coordinate plane, let O be the origin and A, B, C be integer points (whose coordinates are integers). Prove that: for any integer point P inside and on the boundary of $\triangle ABC$, there exist integer points Q, R, which can be the same, inside and on the boundary of $\triangle ABC$ such that $\overrightarrow{OP} = \overrightarrow{OQ} + \overrightarrow{OR}$. (Contributed by Leng Fusheng)

Solution 1   First of all, for integer triangle $\triangle OAC$, whose vertices are all integer points, make an "integer triangulation" satisfying that the vertices of each small triangle in the triangulation are integers and there are no other integers in its interior and on the boundary. The construction is as follows: *select one small triangle in a triangulation at a time, choose a non-vertex integer inside it if there exists, and connect the integer with the three vertices of the small triangle.* After repeating a finite number of times of the operations above, we will get the *"integer triangulation".* Consider the midpoint P' of line segment OP, it is clear that $P' \in \triangle OAC$. So it lies inside some small triangle $\triangle QRS$ in the integer triangulation of $\triangle OAC$. Since $\triangle QRS$ has no integer points except for three vertices, by Pick's theorem the area of $\triangle QRS$ is equal to $\dfrac{1}{2}$.

Consider the homothetic transformation with origin-centered and ratio of 2. Under this transformation, P' becomes P. Denote the images of Q, R, S as Q', R', S', respectively. Thus, P lies inside or on the boundary of $\triangle Q'R'S'$ and the area of $\triangle Q'R'S'$ is 2. Suppose the number of integer points at the boundary and inside of $\triangle Q'R'S'$ are p, q, respectively. From Pick's theorem we know that $\dfrac{p}{2} + q - 1 = 2$. It is easy to know that the midpoints of the three sides of $\triangle Q'R'S'$ are all integer points, and then $p \geq 6$. Combing with the above equation, we get $q \leq 0$. Therefore, there must be $q = 0$, and thus $p = 6$. Since integer point P lies inside or on the boundary of $\triangle Q'R'S'$, thus point P must be one of these 6 integer points on the boundary of $\triangle Q'R'S'$. If P is a vertex of $\triangle Q'R'S'$, say $P = Q'$, then we have $\overrightarrow{OP} = 2\overrightarrow{OQ}$. If P is the midpoint of one side, such as the midpoint of $Q'R'$, then we have $\overrightarrow{OP} = \overrightarrow{OQ} + \overrightarrow{OR}$. In summary, OP can be expressed as the sum of two integer point vectors in $\triangle OAC$.

Solution 2   If P lies on AB, we take $\overrightarrow{OQ} = \overrightarrow{OA}, \overrightarrow{OR} = \overrightarrow{AP}$. If P is on BC, let $\overrightarrow{OQ} = \overrightarrow{OC}, \overrightarrow{OR} = \overrightarrow{CP}$. If P lies on CA, let $Q = P, R = O$. If P lies on the boundary of $\triangle ABC$, then

$$\overrightarrow{OP} = \alpha \cdot \overrightarrow{OA} + \beta \cdot \overrightarrow{OC} (0 < \alpha, \beta < 1, \alpha, \beta \in \mathbb{Q}).$$

Let the least common denominator of α and β be n. For each integer k, denote the fractional parts of $k\alpha$ and $k\beta$ as α_k and β_k, respectively. Thus when $1 \leq k \leq n-1$, there is $(\alpha_k, \beta_k) \neq (0,0)$, while $(\alpha_n, \beta_n) = (0,0)$. Therefore,

$$S = \left\{ A_k \mid \overrightarrow{OA_k} = \alpha_k \cdot \overrightarrow{OA} + \beta_k \cdot \overrightarrow{OC}, 1 \leq k \leq n \right\}$$

is the set of n different integer points inside the parallelogram $OABC$.

If for some $1 \leq k \leq n - 1, A_k \in \triangle ABC$, then $\alpha_k + \beta_k \geq 1$ because the sum of the coefficients of the intersection T of OP and AC expressed in terms of the basis vectors $\overrightarrow{OA}$ and $\overrightarrow{OC}$ is equal to 1. This shows that $0 < \alpha_k, \beta_k < 1$. In particular, both α_k and β_k are not 0. Note that for real number $x \notin \mathbb{Z}$ there is $\{x\} + \{-x\} = 1$. Thus we have $\alpha_k + \alpha_{n-k} = \beta_k + \beta_{n-k} = 1$, i.e., A_{n-k} is a point of symmetry of A_k about the midpoint of AC, and $A_n = O \in \triangle OAC$. And since $A_n = O \in \triangle OAC$, then

$$|S \cap \triangle ABC| < |S \cap \triangle OAC|.$$

Therefore, there exists $1 \leq k \leq \left\lceil \dfrac{n+1}{2} \right\rceil$ such that both A_k and A_{n+1-k} belong to $\triangle OAC$. And there is

$$0 \leq \alpha_k + \beta_k \leq 1, 0 \leq \alpha_{n+1-k} + \beta_{n+1-k} \leq 1.$$

And since P belongs to $\triangle ABC$, there is $1 < \alpha + \beta < 2$, and this is also because the sum of the coefficients of the intersection T of OP and AC expressed in terms of the basis vectors $\overrightarrow{OA}$ and $\overrightarrow{OC}$ is equal to 1. Then

$$(\alpha_k + \alpha_{n+1-k} - \alpha) + (\beta_k + \beta_{n+1-k} - \beta)$$
$$= (\alpha_k + \beta_k) + (\alpha_{n+1-k} + \beta_{n+1-k}) - (\alpha + \beta)$$
$$< 2 - 1$$
$$= 1.$$

Combing $\alpha_k + \alpha_{n+1-k} - \alpha \in \mathbb{Z}$, $\beta_k + \beta_{n+1-k} - \beta \in \mathbb{Z}$, we have

$$(\alpha_k + \alpha_{n+1-k} - \alpha) + (\beta_k + \beta_{n+1-k} - \beta) \leq 0. \qquad (*)$$

On the other hand, since $\alpha, \alpha_k, \alpha_{n+1-k} \in [0, 1)$, there is $\alpha_k + \alpha_{n+1-k} - \alpha \geq 0$. Similarly, there is $\beta_k + \beta_{n+1-k} - \beta \geq 0$. Combing $(*)$ we can find $\alpha_k + \alpha_{n+1-k} = \alpha$, $\beta_k + \beta_{n+1-k} = \beta$. Therefore, we just need to take $Q = A_k, R = A_{n+1-k}$. $\qquad \square$

Second Day
(8:00 – 12:00; August 13, 2019)

⑤ Let prime p divide $2^{2019} - 1$. Define sequence a_n:

$$a_0 = 2, \ a_1 = 1, \ a_{n+1} = a_n + \frac{p^2 - 1}{4} a_{n-1} \ (n \geq 1).$$

Prove that: for any $n \geq 0$, p does not divide $1 + a_n$. (Contributed by Ji Chungang)

Solution   The characteristic equation of sequence a_n is $x^2 - x - \dfrac{p^2 - 1}{4} = 0$,

and its two roots are $\dfrac{1 \pm p}{2}$. Combing $a_0 = 2, a_1 = 1$, we find

$$a_n = \left(\frac{1+p}{2} \right)^n + \left(\frac{1-p}{2} \right)^n.$$

Since $2^{2019} - 1$ is an odd number and $2^{2019} - 1 \equiv (-1)^{2019} - 1 \equiv 1 \pmod 3$,
then $p \neq 2, 3$. Therefore, $p \nmid (1 + a_0)$. For $n \geq 1$, if $p \mid (1 + a_n)$ then by

$$2^n(1 + a_n) = 2^n + (1+p)^n + (1-p)^n \equiv 2^n + 2 \pmod p$$

we have $2^n \equiv -2 \pmod p$. Hence $2^{n-1} \equiv -1 \pmod p$. Combing $2^{2019} \equiv 1 \pmod p$, it follows that

$$-1 \equiv (-1)^{2019} \equiv (2^{n-1})^{2019} \equiv (2^{2019})^{n-1} \equiv 1 \pmod p,$$

and this is a contradiction. Therefore, $p \nmid (1 + a_n)$. $\square$

6   Let $0 \leq x_1 \leq x_2 \leq \cdots \leq x_n \leq 1$, where $n \geq 2$. Prove that

$$\sqrt[n]{x_1 x_2 \cdots x_n} + \sqrt[n]{(1 - x_1)(1 - x_2) \cdots (1 - x_n)} \leq \sqrt[n]{1 - (x_1 - x_n)^2}.$$

(Contributed by Ai Yinghua)

Solution   Denote $a = x_1, b = x_n$. The above inequality obviously holds
when $a = 0, b = 1$. Assume $(a - b)^2 < 1$ and let

$$X = \frac{\sqrt[n]{x_1 \cdots x_n}}{\sqrt[n]{1 - (a - b)^2}}, \quad Y = \frac{\sqrt[n]{(1 - x_1) \cdots (1 - x_n)}}{\sqrt[n]{1 - (a - b)^2}}.$$

By using the inequality of arithmetic and geometric means, we have

$$X = \sqrt[n]{\frac{\sqrt{ab}}{\sqrt{1 - (a - b)^2}} \cdot \frac{\sqrt{ab}}{\sqrt{1 - (a - b)^2}} \cdot x_2 \cdots \cdot x_{n-1}}$$

$$\leq \frac{1}{n} \left(\frac{2\sqrt{ab}}{\sqrt{1 - (a - b)^2}} + x_2 + \cdots + x_{n-1} \right). \qquad \textcircled{1}$$

Similarly, there is

$$Y \leq \frac{1}{n} \left(\frac{2\sqrt{(1 - a)(1 - b)}}{\sqrt{1 - (a - b)^2}} + (1 - x_2) + \cdots + (1 - x_{n-1}) \right). \qquad \textcircled{2}$$

In addition, note that

$$(\sqrt{ab} + \sqrt{(1-a)(1-b)})^2$$
$$= ab + (1-a)(1-b) + 2\sqrt{a(1-a)b(1-b)}$$
$$\leq ab + (1-a)(1-b) + a(1-a) + b(1-b)$$
$$= 1 - (a-b)^2,$$

and then there is

$$\sqrt{ab} + \sqrt{(1-a)(1-b)} \leq \sqrt{1-(a-b)^2}. \qquad ③$$

Combing formulas ①, ②, ③, we obtain the inequality to be proved $X + Y \leq 1$. □

7 As shown in Fig. 7.1, in acute $\triangle ABC$, $AB < AC$. Points D, E, F are on sides BC, CA, AB, respectively, such that points B, C, E, F are concyclic. The symmetric line of BC with respect to line AD intersects ray EF at point K. Let M be the midpoint of EF. Prove that points A, M, P, K are concyclic. (Contributed by Lin Tianqi)

Solution 1 As shown in Fig. 7.2, suppose the midpoint of BC is N. Then the symmetric point N' of N about line AD is on ray DK. From the fact that B, C, E, F are concyclic, we know that $\triangle ABC \sim \triangle AEF$ and $\triangle PBC \sim \triangle PEF$. Since M, N are the midpoints of EF, BC, then

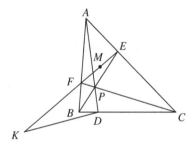

Fig. 7.1

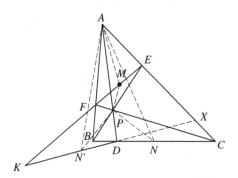

Fig. 7.2

$\triangle ACN \sim \triangle AFM$ and $\triangle BPN \sim \triangle FPM$. Therefore,

$$\angle AN'P = \angle ANP$$

$$= 180° - (\angle ANC + \angle BNP)$$

$$= 180° - (\angle AMF + \angle FMP)$$

$$= 180° - \angle AMP,$$

and thus A, M, P, N' are concyclic.

Let the extension of KD intersect AC at point X. By the given conditions, we know that $\angle ADX = \angle ADB$. Thus

$$\angle K = \angle AEF - \angle AXD$$

$$= \angle ABC - (180° - \angle ADX - \angle XAD)$$

$$= (\angle ABC - 180° + \angle ADB) + \angle CAD$$

$$= \angle CAD - \angle BAD.$$

Combing $\triangle ACN \sim \triangle AFM$, we find

$$\angle N'AM = \angle PAN' + \angle PAM$$

$$= \angle PAN + \angle PAM$$

$$= (\angle CAD - \angle CAN) + (\angle FAM - \angle BAD)$$

$$= \angle CAD - \angle BAD.$$

Therefore, $\angle K = \angle N'AM$, and thus A, M, N', K are concyclic.

Since A, M, P, N' are concyclic and A, M, N', K are concyclic, it follows that A, M, P, N', K are concyclic, and thus the conclusion is valid.

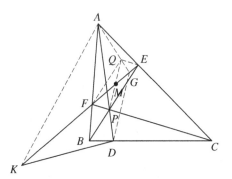

Fig. 7.3

Solution 2   As shown in Fig. 7.3, extend PM to point Q such that $QM = PM$, and then quadrilateral $EPFQ$ is a parallelogram. Extend AQ to EF at point G. Since N, D, I, B are concyclic, we know $\triangle AEF \sim \triangle ABC$. As

$$\angle QEF = \angle PFE = \angle PBC,$$

$$\angle QFE = \angle PEF = \angle PCB,$$

then Q, P are corresponding points in $\triangle AEF \sim \triangle ABC$, and also G, D are corresponding points. Therefore, $\dfrac{AQ}{AG} = \dfrac{AP}{AD}$, and hence $PQ//DG$. From the above conclusion, we know that $\angle ADC = \angle AGF$. Combing the given conditions we get

$$\angle ADK = \angle ADC = \angle AGF = \angle AGK,$$

and hence A, G, D, K are concyclic. Combining $PQ//DG$, we know that

$$\angle APM = \angle ADG = \angle AKG = \angle AKM.$$

Consequently, points A, M, P, K are concyclic and the conclusion is correct.   □

8   Given 8 points in space, among which any 4 points are not coplanar, and between any 2 points there is a line segment marked with a direction connecting them. The resulting graph is called a "directed graph". Starting from any vertex (i.e., the point mentioned above) on a directed graph and moving in a marked direction, if it is impossible to return to the starting vertex, then the graph is called a "good graph". Prove that:

(1) there exists a directed graph such that any changes in the direction of no more than 7 line segments in the graph cannot turn the graph into a good graph;

(2) for any directed graph, it is possible to change the direction of no more than 8 line segments in the graph to turn it into a good graph. (Contributed by Fu Yunhao)

Solution Let the 8 vertices be $A_1, A_2, \ldots, A_8$, respectively.

(1) Consider the following directed graph $A_i \to A_{i+1} \to A_{i+3} \to A_i$, $i = 1, 2, \ldots, 8$, (subscripts should be understood as under modulo 8), and the remaining 4 line segments can be marked in any direction. The above 8 sets of directed line segments do not repeat each other, so after changing the direction of no more than 7 line segments, there is always a set of line segments in the 8 sets whose direction has not been changed. Therefore, the graph is not a good graph.

(2) First, we prove the following lemma.

Lemma *If the 8 vertices are changed to 4 vertices, then a good graph can be obtained by changing the direction of at most 1 line segment.*

Proof of the lemma We may let the 4 vertices be P, Q, R, S. By the pigeonhole principle, there must be a vertex being starting point of at least 2 directed line segment. We may let $P \to Q, P \to R$, and then by symmetry assume $Q \to R$. By the pigeonhole principle, it is easy to know that changing the direction of at most one of $P - S, Q - S, R - S$ can make either all the three line segments point to S or none of them point to S. The graph at this point is a good graph, and we have proved the lemma.

Go back to the original problem. For any $i = 1, 2, \ldots, 8$, define $f(A_i)$ as the number of directed line segments starting from A_i, and it may be useful to assume $f(A_1), f(A_2), \ldots, f(A_8)$ in descending order.

If $f(A_4) \geq 4$, then $f(A_1) + f(A_2) + f(A_3) + f(A_4) \geq 16$, and thus $f(A_5) + f(A_6) + f(A_7) + f(A_8) \leq 12 = C_8^2 - 16$.

If $f(A_4) \leq 3$, then $f(A_5) + f(A_6) + f(A_7) + f(A_8) \leq 4f(A_4) = 12$. Therefore, in either case, there is $f(A_5) + f(A_6) + f(A_7) + f(A_8) \leq 12$. Excluding the $C_4^2 = 6$ line segments inside A_5, A_6, A_7, A_8, there are at most $12 - C_4^2 = 6$ line segments starting from one of A_5, A_6, A_7, A_8 and ending at one of A_1, A_2, A_3, A_4. After changing the direction of these lines, the loops described in the problem can only be found in the lines connecting A_1, A_2, A_3, A_4 or in the lines connecting A_5, A_6, A_7, A_8. Using the lemma for A_1, A_2, A_3, A_4 and A_5, A_6, A_7, A_8, we see that changing the direction of at most 2 line segments so that there is no loop in the lines connecting A_1, A_2, A_3, A_4, and no loop in the lines connecting A_5, A_6, A_7, A_8. Therefore, we only need to change the direction of no more than $6 + 2 = 8$ line segments to turn the original graph into a good graph. $\square$

China Western Mathematical Olympiad

2018 (Haikou, Hainan)

The 2018 China West Mathematical Olympiad (CWMO) was held in Haikou, Hainan Province from August 13th to 18th, hosted by Hainan Middle School in Hainan Province, sponsored by the Organizing Committee of the West China Mathematical Olympiad and Hainan Provincial Mathematics Society.

227 students (89 formal and 138 non-formal contestants) from 23 teams, including the Hong Kong and Macau representative teams, participated in this tournament. In addition, Kazakhstan, Singapore, the Philippines, and other countries also sent teams to participate. The game lasted for two days, with 4 questions a day, each with 15 points, and a perfect score of 120 points. Huang Yizhi from Chengdu No.7 Middle School of the Sichuan team won the first place in individual (105 points). The Sichuan team won the first place in total points. In this competition, 32 students won the gold medals, 29 students won the silver medals, and 25 students won the bronze medals. The score lines that determine the winning ranking of non-formal contestants are the same with that of the official teams: A total of 41 non-formal contestants won the first prize, 49 won the second prize, and 31 won the third prize.

The director of the main test committee of this competition is Leng Gangsong (Professor of Shanghai University). The members are:

Xiong Bin (Professor of East China Normal University);
Feng Zhigang (Shanghai Middle School in Shanghai);
Liu Shixiong (Zhongshan Affiliated Middle School of South China Normal University);
Qu Zhenhua (East China Normal University);
Zou Jin (Gaosi Education);
He Yijie (East China Normal University);
Yang Hu (Beijing Fourth Middle School);
Zhang Duanyang (High School Affiliated to Renmin University of China);
Shi Zehui (Affiliated Middle School of Jilin University);
Zhang Jia (Zhengzhou City First Middle School);
Zhao Yan (The Affiliated Middle School of Harbin Normal University);
Dong Jiaqi (Shanghai Middle School in Shanghai);
Wang Guangting (Shanghai Middle School in Shanghai).

First Day
(8:00 – 12:00; August 15, 2018)

1 Let $x_1, x_2, \ldots, x_{2018}$ be real numbers such that $x_i + x_j \geq (-1)^{i+j}$ for all $1 \leq i < j \leq 2018$. Find the minimum of $\sum_{i=1}^{2018} i x_i$. (Contributed by Shi zehui)

Solution The minimum of $\sum_{i=1}^{2018} i x_i$ is $\dfrac{2037171}{2}$.

On one hand, when $x_1 = x_2 = \cdots = x_{2018} = \dfrac{1}{2}$, we have

$$\sum_{i=1}^{2018} i x_i = \frac{1009 \cdot 2019}{2} = \frac{2037171}{2}.$$

On the other hand, we prove by induction that for all $n(n \geq 7)$, we have

$$\sum_{i=1}^{n} i x_i \geq \frac{n(n+1)}{4}.$$

When $n = 7$, there are

$$2(x_2 + x_6) \geq 2, 4(x_4 + x_6) \geq 4, x_1 + x_7 \geq 1,$$

$$x_3 + x_5 \geq 1, 2(x_3 + x_7) \geq 2, 4(x_5 + x_7) \geq 4,$$

and taking the sum implies the conclusion.

Supposing the conclusion already holds for $n(n \geq 7)$, consider the case for $n+1$. Let $y_i = x_{i+1}$ $(i = 1, 2, \ldots, n)$. Then for $1 \leq i < j \leq n$ we have

$$y_i + y_j = x_{i-1} + x_{j-1} \geq (-1)^{(i+1)+(j+1)} = (-1)^{i+j}.$$

Hence by induction hypothesis it follows that $\sum_{i=1}^{n} iy_i \geq \dfrac{n(n+1)}{4}$, so that

$$\sum_{i=1}^{n+1} (i-1)x_i \geq \frac{n(n+1)}{4}. \qquad \text{(1)}$$

On the other hand, we have

$$x_i + x_{i+2} \geq 1 \ (i = 1, 2, \ldots, n+1, x_{n+2} = x_1, x_{n+3} = x_2),$$

where taking the sum we obtain $2 \sum_{i=1}^{n+1} x_i \geq n+1$, or equivalently,

$$\sum_{i=1}^{n+1} x_i \geq \frac{n+1}{2}. \qquad \text{(2)}$$

Therefore, by adding (1) and (2) we get

$$\sum_{i=1}^{n+1} ix_i \geq \frac{n(n+1)}{4} + \frac{n+1}{2} = \frac{(n+1)(n+2)}{4}.$$

Thus the induction is complete and the general conclusion holds. $\quad\square$

2 Given integer $n \geq 2$, suppose $x_1, x_2, \ldots, x_n$ are positive real numbers such that $x_1 x_2 \ldots x_n = 1$. Prove that

$$\{x_1\} + \{x_2\} + \cdots + \{x_n\} < \frac{2n-1}{2}.$$

Here $\{x\}$ means the fractional part of a real number x.
(Contributed by Wang Guangting)

Solution We first prove a lemma.

Lemma *For any two positive real numbers α, β, if $\alpha\beta < 1$, then $\{\alpha\} + \{\beta\} < 1 + \alpha\beta$.*

Proof of lemma Since $(1 - \{\alpha\})(1 - \{\beta\}) > 0$, we get $1 + \{\alpha\}\{\beta\} > \{\alpha\} + \{\beta\}$. Then by the condition we have $\{\alpha\}\{\beta\} < \alpha\beta = \{\alpha\beta\}$. Hence $\{\alpha\} + \{\beta\} < 1 + \alpha\beta$. Thus, the lemma is proven.

Returning to the original problem, we apply induction to prove the conclusion.

When $n = 2$, we assume without loss of generality that $x_1 \geq x_2$. Then since $x_1 x_2 = 1$, we have $x_1 \geq 1, x_2 \leq 1$. If $x_2 = 1$, then $x_1 = x_2 = 1$, and the result holds. If $x_2 < 1$, then $[x_2] = 0$. Let $[x_1] = s$, then $x_1 < s + 1$. Hence

$$\{x_1\} + \{x_2\} = x_1 + x_2 - s = x_1 + \frac{1}{x_1} - s < (s+1) + \frac{1}{s+1} - s$$

$$= 1 + \frac{1}{s+1} \leq 1 + \frac{1}{2} = \frac{3}{2}.$$

Supposing the conclusion holds for $n = k$ $(k \geq 2, k \in \mathbb{N}^*)$, next we consider the case when $n = k + 1$.

If $x_{i_1} x_{i_2} \geq 1$ for all i_1, i_2 $(1 \leq i_1 < i_2 \leq n)$, then $x_1 = x_2 = \cdots = x_n = 1$, and

$$\{x_1\} + \{x_2\} + \cdots + \{x_n\} = 0.$$

If $x_{i_1} x_{i_2} < 1$ for some i_1, i_2 $(1 \leq i_1 < i_2 \leq n)$, then without loss of generality assume that $i_1 = n - 1, i_2 = n$. From the lemma we have

$$\sum_{i=1}^{n} \{x_i\} = \sum_{i=1}^{n-2} \{x_i\} + \{x_{n-1}\} + \{x_n\}$$

$$< \sum_{i=1}^{n-2} \{x_i\} + (x_{n-1}x_n + 1)$$

$$= \sum_{i=1}^{n-2} \{x_i\} + \{x_{n-1}x_n\} + 1$$

$$< \frac{2(n-1) - 1}{2} + 1$$

$$= \frac{2n - 1}{2}.$$

Here the first inequality follows from the lemma and the second follows from the induction hypothesis. Therefore, the conclusion holds when $n = k + 1$. The proof is complete. $\square$

3   Let set $M = \{1, 2, \ldots, 10\}$, and T be a set family consisting of some 2-element subsets of M, such that for any two elements $\{a, b\}, \{x, y\}$ in T, $(ax + by)(ay + bx)$ is not divisible by 11. Find the maximum number of elements in T. (Contributed by Liu Shixiong)

Solution 1 The maximum number of elements in T is 25.

Since the elements of M are all coprime to 11, for any 2-element subset $\{a, b\}$ of M, there is a unique pair (k, l) such that $ka \equiv b \pmod{11}, lb \equiv a \pmod{11}$.

Define $f(\{a, b\}) = \{k, l\}$, where k, l are chosen based on the relations above. If $k = l$, we define $f(\{a, b\}) = \{k\}$. Then from the definition we have $kla \equiv a \pmod{11}$, so $kl \equiv 1 \pmod{11}$.

Hence for any $\{a, b\} \in T$, we have $f(\{a, b\}) \in \{\{2, 6\}, \{3, 4\}, \{5, 9\}, \{7, 8\}, \{10\}\}$.

Let $M_1 = \{\{a, b\} | a, b \in M, f(\{a, b\}) = \{2, 6\}\}$, $M_2 = \{\{a, b\} | a, b \in M, f(\{a, b\}) = \{3, 4\}\}$, $M_3 = \{\{a, b\} | a, b \in M, f(\{a, b\}) = \{5, 9\}\}$, $M_4 = \{\{a, b\} | a, b \in M, f(\{a, b\}) = \{7, 8\}\}$, $M_5 = \{\{a, b\} | a, b \in M, f(\{a, b\}) = \{10\}\}$.

We consider the pairs $\{a, ka\}$ ($a = 1, 2, \ldots, 10$) in the sense of $\mod 11$. It is easy to verify that for $2 \le k \le 9$ we have $k^2 \not\equiv 1 \pmod{11}$, so the 10 pairs above are different from each other, and hence each of M_1, M_2, M_3, M_4 has 10 elements. Set M_5, on the other hand, has 5 elements.

Now for any $\{a, b\} \in M_1, \{x, y\} \in M_3$, assume $2a \equiv b \pmod{11}$, $5x \equiv y \pmod{11}$, then $ax + by \equiv ax + 10ax \equiv 0 \pmod{11}$, which means they cannot both belong to T. Hence at least one of $T \cap M_1$ and $T \cap M_3$ is empty. Similarly at least one of $T \cap M_2$ and $T \cap M_4$ is empty. Therefore T has at most $10 + 10 + 5 = 25$ elements.

On the other hand, let $T = M_1 \cup M_3 \cup M_5$, then T has 25 elements. For any two elements of T, let them be $\{a, b\}, \{x, y\}$, and assume $ka \equiv b \pmod{11}, lx \equiv y \pmod{11}$, where $k, l \in \{2, 3, 10\}$, then $ax + by \equiv (kl + 1)ax \pmod{11}$, $ay + bx \equiv (k + l)ax \pmod{11}$. It is simple to verify that neither $kl + 1$ nor $k + l$ can be a multiple of 11, so $(ax + by)(ay + bx)$ is not divisible by 11, i.e., T satisfies the conditions.

In summary, the maximum number of elements in T is 25. $\qquad\Box$

Solution 2 The maximum number of elements in T is 25.

From the given condition it follows that for any $a, b \in M$ satisfying $a \ne b, a \ne 11 - b$, the two sets $\{a, b\}, \{a, 11 - b\}$ cannot both belong to T. Hence for any $a \in M$, we can only choose 5 out of the 9 two-element subsets of M containing a, so $|T| \le \dfrac{5 \times 10}{2} = 25$.

On the other hand, let $T = \{\{a, 2a\} | a \in M\} \cup \{\{a, 3a\} | a \in M\} \cup \{\{a, 10a\} | a \in M, a \le 5\}$, which is similar to the construction in Solution 1, and we can prove that T satisfies the conditions and $|T| = 25$. $\qquad\Box$

4 As shown in Fig. 4.1, in an acute-angled triangle ABC, $AB > AC$, the points E, F lie on sides AC, AB, respectively, such that $BF + CE = BC$. Let I_B, I_C be the escenters of $\triangle ABC$ inside $\angle B, \angle C$, respectively, and lines EI_C, FI_B intersect at T. Let K be the midpoint of arc $\overset{\frown}{BAC}$, and line KT intersects the circumcircle of $\triangle ABC$ at K, P. Prove that T, F, P, E are concyclic. (Contributed by Zhang Jia)

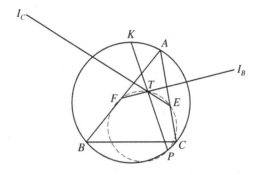

Fig. 4.1

Solution As shown in Fig. 4.2, connecting AK, AI_B, AI_C, since all of AK, AI_B, AI_C are the outer angle bisector of $\angle BAC$, it follows that I_B, A, K, I_C are collinear. From the property of the escenter we know that K is the midpoint of $I_B I_C$, and $KI_B = KI_C = KB$.

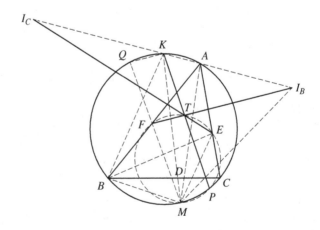

Fig. 4.2

Choose a point D on side BC, such that $BD = BF$. Since $BF + CE = BC = BD + CD$, we have $CD = CE$. Let M be the second intersection point of line KD and the circumcircle of $\triangle ABC$, and let Q be the second intersection point of line MF and the circumcircle of $\triangle ABC$. Connecting $BK, BM, I_B M, MC, QK$, note that K is the midpoint of $\overset{\frown}{BAC}$, so we have $\angle KBD = \angle KMB$. Hence, $\triangle KBD \sim \triangle KMB$, and consequently $\dfrac{KB}{BD} = \dfrac{MK}{MB}$. Also since $KB = KI_B$ and $BD = BF$, we have $\dfrac{MK}{KI_2} = \dfrac{MB}{BF}$, and combining with $\angle I_B KM = \angle AKM = \angle ABM = \angle FBM$, we obtain that $\triangle I_B KM \sim \triangle FBM$. Thus $\angle BFM = \angle KI_B M = \angle AI_B M$, and A, F, M, I_B are concyclic.

Further we have $\angle I_C KQ = \angle AMQ = \angle AMF = \angle AI_B F$, and hence $QK \parallel I_B F$. Therefore, $\angle FMP = \angle QMP = 180° - \angle QKP = 180° - \angle QKT = \angle KTF$, from which we see that F, T, P, M are concyclic. By the same method we may prove that E, T, M, P are concyclic, so E, T, F, M, P all lie on the same circle. $\qquad\square$

Second Day
(8:00 – 12:00; August 16, 2018)

5 As shown in Fig. 5.1, in acute triangle ABC, $AB < AC$, O is its circumcenter, and M is the midpoint of BC. Let the circumcircle of $\triangle AOM$ intersect the extension of segment AB at D, and intersect segment AC at E. Prove that $DM = EC$. (Contributed by He Yijie)

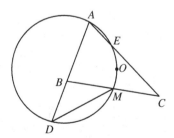

Fig. 5.1

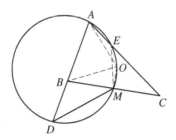

Fig. 5.2

Solution　Connect EM, OA, OB, OM. From the conditions, we have $BM = MC$, so $\angle BDM = \angle ADM = \angle MEC$. Since $OM \perp BC$, we also have

$$\angle DMB = \angle DMO - \angle BMO = (180° - \angle OAD) - 90°$$

$$= 90° - \angle OAB = \frac{1}{2} \angle AOB = \angle ACB = \angle ECM.$$

Hence $\triangle BDM \cong \triangle MEC$, and $DM = EC$.

6　Let $n \geq 2$ be an integer, and $a_1 \geq a_2 \geq \cdots \geq a_n$ be positive real numbers. Prove that

$$\left(\sum_{i=1}^{n} \frac{a_i}{a_{i+1}} \right) - n \leq \frac{1}{2a_1 a_n} \sum_{i=1}^{n} (a_i - a_{i+1})^2,$$

where $a_{n+1} = a_1$. (Contributed by Zhang Duanyang)

Solution　Since $a_{i+1} \geq a_i$ for $1 \leq i \leq n-1$, it follows that

$$\left(\sum_{i=1}^{n} \frac{a_i}{a_{i+1}} \right) - n = \sum_{i=1}^{n-1} \left(\frac{a_i}{a_{i+1}} - 1 \right) - \left(1 - \frac{a_n}{a_1} \right)$$

$$= \left(\sum_{i=1}^{n-1} \frac{a_i - a_{i+1}}{a_{i+1}} \right) - \left(\sum_{i=1}^{n-1} \frac{a_i - a_{i+1}}{a_i} \right)$$

$$= \sum_{i=1}^{n-1} \frac{(a_i - a_{i+1})(a_1 - a_{i+1})}{a_1 a_{i+1}}$$

$$\leq \frac{1}{a_1 a_n} \sum_{i=1}^{n-1} (a_i - a_{i+1})(a_1 - a_{i+1}).$$

Further, we have

$$\sum_{i=1}^{n-1} (a_i - a_{i+1})(a_1 - a_{i+1})$$

$$= \sum_{i=1}^{n-1} (a_i - a_{i+1})a_1 - \sum_{i=1}^{n-1} (a_i - a_{i+1})a_{i+1}$$

$$= (a_1 - a_n)a_1 - \sum_{i=1}^{n-1} a_i a_{i+1} + \sum_{i=2}^{n} a_i^2 = \sum_{i=1}^{n} a_i^2 - \sum_{i=1}^{n} a_i a_{i+1}$$

$$= \frac{1}{2} \sum_{i=1}^{n} (a_i - a_{i+1})^2.$$

Therefore, the desired result is proven. □

7 Let p, c be a prime number and a composite number, respectively. Prove that there exist positive integers m, n, such that

$$0 < m - n < \frac{[n+1, n+2, \ldots, m]}{[n, n+1, \ldots, m-1]} = p^c.$$

Here $[x_1, x_2, \ldots, x_k]$ represents the least common multiple of positive integers $x_1, x_2, \ldots, x_k$. (Contributed by He Yijie)

Solution Let $m = p^{2c-1}, n = p^{2c-1} - p^{c-1}$. We shall prove that (m, n) is a valid pair. Apparently $0 < m - n = p^{c-1} < p^c$, so it suffices to show that $\frac{[L, m]}{[n, L]} = p^c$, where $L = [n+1, n+2, \ldots, m-1]$. We use $v_r(l)$ to denote the power of a prime number r in the factorization of positive integer l.

Note that for $m-i$ $(i = 1, 2, \ldots, p^c - 1)$, we have $v_p(m-i) = v_p(i) < c-1$, so

$$v_p([L, p^{2c-1}]) = 2c - 1, \quad v_p([p^{c-1}(p^c - 1), L]) = c - 1.$$

Hence,

$$v_p\left(\frac{[L, m]}{[n, L]}\right) = v_p([L, p^{2c-1}]) - v_p([p^{c-1}(p^c - 1), L]) = c. \qquad \text{①}$$

Next we show that for any prime number $q \neq p$, we have

$$v_q(n) \leq v_q(L). \qquad \text{②}$$

Since c is composite, we may write

$$c = st(s \geq t \geq 2), \text{ then } p^c - 1 = (p^s - 1)(1 + p^s + p^{2s} + \cdots + p^{(t-1)s}).$$

Let $x = p^s - 1, y = 1 + p^s + p^{2s} + \cdots + p^{(t-1)s}, d = \gcd(x, y)$, then we have

$$xd < yd < p^{c-1}.$$

In fact, it is obvious that $x < y$, and from $y \equiv t \pmod{x}$ we get $d = \gcd(x, t) \leq t$. If $p = t = 2$, then since $d = (2^s - 1, 2) = 1$, we have $yd = y = 1 + 2^s < 2^{2s-1} = 2^{c-1}$. If p, t are not both 2, then $x \geq p^t - 1 \geq pt \geq pd$, and $yd \leq \dfrac{xy}{p} = \dfrac{p^c - 1}{p} < p^{c-1}$.

Thus, $n < n + xd < n + yd < n + p^{c-1} = m$, so that

$$\max\{v_q(n + xd), v_q(n + yd)\} \leq v_q(L).$$

Let $v_q(d) = v$. If $v_q(x) \geq v_q(y)$, then $v_q(y) = v$, and combining with $n = p^{c-1}xy$ we obtain that

$$v_q(n + xd) = v_q(p^{c-1}xy + xd) = v_q(x) + v_q(p^{c-1}y + d)$$

$$\geq v_q(x) + v = v_q(x) + v_q(y) = v_q(n).$$

If $v_q(x) \leq v_q(y)$, similarly we may get $v_q(n + yd) \geq v_q(n)$. Therefore, $v_q(n) \leq \max\{v_q(n + xd), v_q(n + yd)\} \leq v_q(L)$, and the inequality ② follows.

Now we see that for any prime number $q \neq p$, we have $v_q([n, L]) = v_q(L)$, so

$$v_q\left(\frac{[L, m]}{[n, L]}\right) = v_q([L, p^{2c-1}]) - v_q([n, L]) = v_q(L) - v_q(L) = 0. \qquad \text{(3)}$$

Combining (1) and (3), we conclude that $\dfrac{[L, m]}{[n, L]} = p^c$, and pair (m, n) meets all the requirements. $\qquad\square$

Remark When constructing m and n, we need to concern both the power of p and the powers of other primes in the expression $\dfrac{[n+1, n+2, \ldots, m]}{[n, n+1, \ldots, m-1]}$, as well as keeping $m - n$ small. This stumps many contestants as they failed to attend to everything at one time. After finding the construction $m = p^{2c-1}, n = p^{2c-1} - p^{c-1}$, we may use the assumption that c is composite to factorize $p^c - 1$, in order to show that $p^c - 1$ divides L. However, some participants neglected the case where $x = p^s - 1$ and $y = 1 + p^s + p^{2s} + \cdots + p^{(t-1)s}$ are not coprime, and claimed that x, y both dividing L implies xy dividing L, which is wrong.

8 Let n, k be given positive integers, where n is even, $k \geq 2$ and $n > 4k$. Suppose there are n points on the circumference of a circle, and we shall call a set of $\dfrac{n}{2}$ chords of the circle a "matching" if they do not intersect inside the circle, and their endpoints are exactly the n chosen points. Find the maximum integer m, such that for any "matching", it is possible to find k consecutive points (among the chosen points), which contain both endpoints of at least m chords. (Contributed by Qu Zhenhua)

Solution The answer is $\left\lceil \dfrac{k}{3} \right\rceil$. (Here $\lceil x \rceil$ means the smallest integer that is greater than or equal to x). Let $m = \left\lceil \dfrac{k}{3} \right\rceil$. Then it is easy to see that $k \geq 2m$, and $3m - 2 \leq k \leq 3m$.

We first give an example of matching, where any k consecutive points contain the endpoints of at most m chords.

Without loss of generality, we assume that the chosen points are the vertices of a regular n-gon. Since n is even, and $n > 4k \geq 8m$, we may divide the n points into 6 consecutive pieces $M_1, M_2, X_1, M_3, M_4, X_2$, which contain $2m, 2m, x, 2m, 2m, x$ points, respectively. In each M_i, we pair up one point at the head and one point at the tail, thus constructing m chords which are parallel to each other. Between X_1 and X_2, we may also construct x parallel chords, which we call "long chords". All these chords are non-intersecting within the circle, so they compose a matching. Let K be a set of k consecutive points, then K may not contain both endpoints of a "long chord", since otherwise K should contain at least $4m + 2 > k$ points. If K contains both endpoints of a chord in M_i, then K must contain the two middle points of M_i. Note that this means K cannot contain chords in three different M_i's, since otherwise K should contain the two middle points in such as M_1, M_3, respectively, and consequently $|K| > \dfrac{n}{2} > k$, leading to a contradiction. Hence K contains chords in at most two different M_i's. If K contains chords in only one M_i, then K contains at most m chords. On the other hand, if K contains chords in two M_i's, let them be M_i, M_{i+1}. Then $|K \cap M_i| = s, |K \cap M_{i+1}| = t$, where $s + t \leq k$. Since K consists of k consecutive points, the points in $|K \cap M_i|$ are the last s points in M_i, while the points in $|K \cap M_{i+1}|$ are the first t points in M_{i+1}, and $s > m, t > m$. This implies that K contains exactly $(s - m) + (t - m)$ chords, and in this case

$$(s - m) + (t - m) = s + t - 2m \leq k - 2m \leq 3m - 2m = m.$$

Next we show that for any matching, there are always k consecutive points that contain the endpoints of at least m chords.

For a chord in the matching, define its length to be the number of points on the minor arc divided by the chord (including both endpoints). Let the n points be $P_1, P_2, \ldots, P_n$, in the clockwise order.

Case 1: Some chord in the matching has length $l \in [2m, k]$. In this case we may choose k consecutive points, containing the two endpoints of this "long chord" and all the l points on the minor arc. Since all these chords are non-intersecting, the points can only match the points on the same side of the chord, so there are at least $\dfrac{l}{2} \geq m$ chords.

Case 2: No chord in the matching has length in interval $[2m, k]$, but some chord has length greater than k. Now we choose the chord with the smallest length among all the chords whose lengths are greater than k. Without loss of generality let it be $P_1 P_l (l > k)$, and we take points $P_2, P_3, \ldots, P_{k+1}$, and claim that there are at least m chords among them.

In fact, we divide $P_2, P_3, \ldots, P_{l-1}$ into several pieces in the following way: suppose P_2 is connected to P_i in the matching, and the points from P_2 to P_i are called the first piece; suppose P_{i+1} is connected to P_j, and the points from P_{i+1} to P_j are called the second piece; so on and so forth. Suppose we finally obtain s pieces, and their lengths are $l_1, l_2, \ldots, l_s$, respectively. From $l_i < l$, together with the assumption on l, we find that $l_i \leq 2m - 2$. Hence, we have

$$l_1 + l_2 + \cdots + l_s = l - 2 \geq k - 1 \geq 2m - 1.$$

Thus, there exists unique positive integer i with $1 \leq i < s$, such that

$$l_1 + l_2 + \cdots + l_i \leq 2m - 2 < l_1 + l_2 + \cdots + l_{i+1}.$$

Let $l_1 + l_2 + \cdots + l_i = 2m - 2t, l_{i+1} = 2p$, then $t < p \leq m - 1$. Since

$$k - (2m - 2t) \geq p + t \iff k + t \geq p + 2m \Longleftarrow k + 1 \geq 3m - 1,$$

then $P_2, P_3, \ldots, P_{k+1}$ contain not only all the points in the first i pieces but also at least $p + t$ points in the $(i + 1)$-th piece. The first i pieces contain exactly $m - t$ chords. On the other hand, choosing $p + t$ points out of p pairs admits at least t matching pairs, so we have included the endpoints of at least t chords in the $(i + 1)$-th piece. Therefore, points $P_2, P_3, \ldots, P_{k+1}$ contain at least m chords.

Case 3: All chords have lengths smaller than $2m$. We choose the longest chord, and assume it to be $P_1 P_{l_1}, l_1 < 2m$. Then consider the chord $P_{l_i+1} P_j$. By maximality of l_1, chord $P_1 P_{l_1}$ cannot be located on one side of $P_{l_i+1} P_j$, so $j = l_1 + l_2, l_2 < 2m$. Then we consider the chord with P_{j+1} as an endpoint. Continuing this process, we may divide the n points into several pieces, with each piece containing fewer than $2m$ points, and the first and last points in each piece are connected by a chord. Similar to the proof in Case 2, we can take k consecutive points starting at P_1, and there are at least m chords. $\qquad\qquad\square$

China Western Mathematical Olympiad

2019 (Zunyi, Guizhou)

The 2019 China West Mathematical Olympiad (CWMO) was held in Zunyi, Guizhou Province from August 10th to 15th, sponsored by the Organizing Committee of the West China Mathematical Olympiad and Guizhou Provincial Mathematics Society, and hosted by Zunyi No.4 Middle School in guizhou Province,

219 students (85 formal and 134 non-formal contestants) from 23 teams. In addition, the Philippines, Kazakhstan, Singapore, and other countries and regions also sent teams to participate in the competition. The game lasted for two days, with 4 questions a day, each with 15 points, and a perfect score of 120 points. Li Yanjie from Dexing City Tongkuang Middle School of the Jiangxi team won the first place in individual. The Sichuan team won the first place in total points. In this competition, 33 students won the gold medals, 33 students won the silver medals, and 19 students won the bronze medals. The score lines that determine the winning ranking of non-formal contestants are the same with that of the official contestants: 36 non-formal contestants won the first prize, 36 won the second prize, and 55 won the third prize.

The director of the main test committee of this competition is Leng Gangsong (Professor of Shanghai University). The members are:

Xiong Bin (Professor of East China Normal University);
Liu Shixiong (Zhongshan Affiliated Middle School of South China Normal University);
Feng Zhigang (Shanghai Middle School in Shanghai);
Qu Zhenhua (East China Normal University);
Zou Jin (Gaosi Education);
He Yijie (East China Normal University);
Zhang Duanyang (High School Affiliated to Renmin University of China);
Zhang Jia (Zhengzhou City First Middle School);
Shi Zehui (Affiliated Middle School of Jilin University);
Yang Mingliang (Leqing Zhilin Middle School in Zejiang Province);
Luo Zhenhua (East China Normal University);
Dong Jiaqi (Shanghai Middle School in Shanghai);
Wang Guangting (Shanghai High School in Shanghai);
Zhou Tianyou (Shanghai High School in Shanghai).

First Day
(8:30 – 12:30; August 13, 2019)

1 Find all positive integer n, such that $3^n + n^2 + 2019$ is a perfect square. (Contributed By Zou Jin)

Solution   If n is odd, then $3^n + n^2 + 2019 \equiv (-1)^n + n^2 + 3 \equiv 3 \pmod 4$, which means it cannot be a square number. Hence n must be even.

Let $n = 2m \ (m \in \mathbb{N}^*)$. Then

$$3^n + n^2 + 2019 = 3^{2m} + 4m^2 + 2019 > (3^m)^2.$$

Thus,

$$3^{2m} + 4m^2 + 2019 \geq (3^m + 1)^2,$$

which means $3^m - 2m^2 \leq 1009$.

We will prove by induction that $3^m > 2m^2 + 1009$, for $m \geq 7$. The assertion is true when $m = 7$. Assuming it is true for m, we have

$$3^{m+1} > 3(2m^2 + 2019) = 6m^2 + 3027 > 2(m + 1)^2 + 1009.$$

Then the assertion is true for $m + 1$. The inequality is proven.

Now we consider $m \leq 6$.

When $m = 1$, $3^{2m} + 4m^2 + 2019 = 2032$, not a square.

When $m = 2$, $3^{2m} + 4m^2 + 2019 = 2116 = 46^2$, a perfect square.

When $m = 3$ or $m = 6$, $3^{2m} + 4m^2 + 2019 \equiv 3 \pmod 9$, not a square.
When $m = 4$, $3^{2m} + 4m^2 + 2019 \equiv 6 \pmod 7$, not a square.
When $m = 5$, $3^{2m} + 4m^2 + 2019 \equiv 3 \pmod 5$, not a square.
In summary, the only possible case is $n = 4$. □

2 As shown in Fig. 2.1, in acute triangle ABC, $AB > AC$, and O, H are its circumcenter and orthocenter, respectively. Let M be the midpoint of BC. Suppose the extension of AM intersects the circumcircle of $\triangle BHC$ at K, and line HK intersects BC at N. Prove that if $\angle BAM = \angle CAN$, then $AN \perp OH$. (Contributed by Zhang Jia)

Solution 1 As Shown in Fig. 2.2, extend AM to K' such that $MK' = AM$. Then $ABK'C$ is a parallelogram. Since $BH \perp AC, BK' \parallel AC$, we have $BH \perp BK'$. Similarly, $CK' \perp CH$. So B, H, C, K' are concyclic, which means K' coincides with K, and KH is the diameter of this circle.

Let T be the midpoint of KH, then T is the circumcenter of $\triangle BHC$, and the circumcircle has the same radius as the circumcircle of $\triangle ABC$. Hence O, T are symmetric with respect to BC, so that $AH \parallel OT$ and $AH = OT$, and $AOTH$ is a parallelogram. Thus we have $AO \parallel KH$, and $\angle OAM = \angle AKH$.

Since AM, AN are isogonal in $\angle BAC$, and AO, AH are also isogonal in $\angle BAC$, it follows that $\angle OAM = \angle HAN$, which implies $\angle HAN = \angle AKH$. Combining with $\angle AHK = \angle NHA$, we have $\triangle AHK \sim \triangle NHA$, so

$$AH^2 = HN \cdot HK = HN^2 + HN \cdot NK.$$

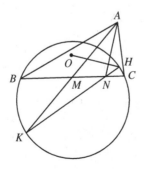

Fig. 2.1

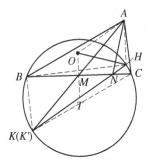

Fig. 2.2

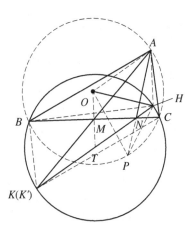

Fig. 2.3

By the circle-power theorem we have $AH^2 - HN^2 = HN \cdot NK = BN \cdot NC = AO^2 - ON^2$, and therefore $AN \perp OH$. □

Solution 2 As shown in Fig. 2.3, similar to solution I, we obtain $\angle HAN = \angle AKH$. Now let P be the second intersection of AN and $\odot O$, and connect PO, PH. Then by the circle-power theorem we have $AN \cdot NP = BN \cdot NC = HN \cdot NK$, so that A, H, P, K are concyclic.

Hence $\angle APH = \angle AKH = \angle OAM = \angle HAN = \angle HAP$, and $AH = HP$. Then combining with $OA = OP$, we see that OH is the perpendicular bisector of segment AP. Therefore, $AN \perp OH$. □

3 Let $S = \{(i, j) \mid i, j = 1, 2, \ldots, 100\}$ be a set consisting of 100×100 grid points on the coordinate plane. We color the points in S with 4 different colors. Find the maximum number of rectangles, whose vertices are points in S with different colors, and whose sides are parallel to the coordinate axes.
(Contributed by Qu Zhenhua)

Solution The maximum is 9375000.

Let A, B, C, D denote the four given colors, $R_i = \{(x, i) \mid x = 1, 2, \ldots, 100\}$ be the i-th row of the grid array, and $T_j = \{(j, y) \mid y = 1, 2, \ldots, 100\}$ be the j-th column of the grid array. We call a rectangle with the desired properties (i.e. its vertices are points in S with different colors and its sides are parallel to the coordinate axes) a "colorful" rectangle.

First we prove the following lemma.

Lemma *If there are k columns in S, such that the corresponding two points of rows R_i, R_j on each column have the same color, then the number of "colorful" rectangles with vertices in these two rows is not greater than $\dfrac{(100 - k)^2}{4}$.*

Proof of lemma If the corresponding two points of R_i, R_j on a column have the same color, then this point pair cannot be vertices of the same colorful rectangle. If $(x, i), (x, j)$ have colors A, B, then they form a colorful rectangle with $(y, i), (y, j)$ if and only if the latter point pair have colors C, D.

Among 100 pairs $(x, i), (x, j)$ $(1 \leq x \leq 100)$, suppose there are l_1 pairs with colors A, B, l_2 pairs with colors A, C, l_3 pairs with colors A, D, l_4 pairs with colors B, C, l_5 pairs with colors B, D, and l_6 pairs with colors C, D. Then $l_1 + l_2 + l_3 + l_4 + l_5 + l_6 = 100 - k$.

Note that $R_i R_j$ generate exactly $l_1 l_6 + l_2 l_5 + l_3 l_4$ colorful rectangles, and

$$l_1 l_6 + l_2 l_5 + l_3 l_4 \leq \frac{(l_1 + l_6)^2}{4} + \frac{(l_2 + l_5)^2}{4} + \frac{(l_3 + l_4)^2}{4}$$

$$\leq \frac{(l_1 + l_2 + l_3 + l_4 + l_5 + l_6)^2}{4}$$

$$= \frac{(100 - k)^2}{4}$$

Hence the lemma is proven.

Now come back to the original problem. Suppose there are $k_{i,j}$ columns, such that the corresponding point pair of rows $R_i R_j$ on each column have the same color, then $\sum\limits_{1 \le i < j \le 100} k_{i,j}$ is the number of pairs of points which have the same x-coordinate and the same color. Then we calculate this number by column. Let there be $a_j b_j, c_j, d_j$ points in T_j with colors A, B, C, D, respectively ($a_j + b_j + c_j + d_j = 100$). Then there are exactly $\binom{a_j}{2} + \binom{b_j}{2} + \binom{c_j}{2} + \binom{d_j}{2}$ pairs of monochromatic pairs in T_j. Thus

$$
\sum_{1 \le i < j \le 100} k_{i,j} = \sum_{j=1}^{100} \binom{a_j}{2} + \binom{b_j}{2} + \binom{c_j}{2} + \binom{d_j}{2}
$$

$$
= \sum_{j=1}^{100} \left[\frac{1}{2}(a_j^2 + b_j^2 + c_j^2 + d_j^2) - \frac{1}{2}(a_j + b_j + c_j + d_j) \right]
$$

$$
\ge \sum_{j=1}^{100} \left[\frac{1}{8}(a_j + b_j + c_j + d_j)^2 - \frac{1}{2}(a_j + b_j + c_j + d_j) \right]
$$

$$
= 100 \cdot \left(\frac{1}{8} \cdot 100^2 - \frac{1}{2} \cdot 100 \right)
$$

$$
= 120000.
$$

Combining the lemma and the inequality above we conclude that the number of colorful rectangles cannot exceed

$$
\sum_{1 \le i < j \le 100} \frac{(100 - k_{i,j})^2}{4} \le \sum_{1 \le i < j \le 100} \frac{100 \cdot (100 - k_{i,j})}{4}
$$

$$
\le 2500 \cdot \binom{100}{2} - 25 \sum_{1 \le i < j \le 100} k_{i,j}
$$

$$
\le 9375000.
$$

Finally, we construct an example of coloring where there are exactly 9375000 colorful rectangles. We divide the 100×100 grid into 16 regions, each with size 25×25, and color each region with the same color, as shown in the figure below.

A	B	C	D
B	A	D	C
C	D	A	B
D	C	B	A

In this coloring method, when the first point of two rows belong to the same region, they do not generate any colorful rectangle; when the first point of two rows belong to different regions, there are four pairs of regions in these two rows, with each pair generating 25×25 colorful rectangles. Since there are $25 \times 25 \times \binom{4}{2}$ ways two choose two rows in different regions, the total number of colorful rectangles is $4 \times 25 \times 25 \times \left(25 \times 25 \times \binom{4}{2} \right) = 9375000$.

In summary, the maximal number of colorful rectangles is 9375000. $\square$

Remark   During the contest many students assumed the wrong answer (which is 6250000). The starting point of the problem is the construction, where we divide the grid into 4×4 regions, and color appropriately to obtain 9375000 colorful rectangles. The proof is then based on the estimation of the number of monochromatic pairs. The lemma gives an estimation of the number of colorful rectangles in each pair of rows. By calculating twice and Cauchy's inequality we can get a lower bound of $\sum_{1 \leq i < j \leq 100} k_{i,j}$, and combining with the lemma and appropriate inequality.

4   Let $n(n \geq 2)$ be a given integer. Find the minimal real number λ, such that for any real numbers $x_1, x_2, \ldots, x_n \in [0,1]$, there exist $\varepsilon_1, \varepsilon_2, \ldots, \varepsilon_n \in \{0,1\}$, such that the following inequality holds for any $1 \leq i < j \leq n$:

$$\left| \sum_{k=i}^{j} (\varepsilon_k - x_k) \right| \leq \lambda.$$

(Contributed by Zhang Duanyang)

Solution   The minimum of λ is $\dfrac{n}{n+1}$.

On one hand, we choose $x_1 = x_2 = \cdots = x_n = \dfrac{1}{n+1}$. For any $\varepsilon_1, \varepsilon_2, \ldots, \varepsilon_n \in \{0, 1\}$, we show that there exist $1 \le i < j \le n$ such that $|\sum_{k=i}^{j}(\varepsilon_k - x_k)| \ge \dfrac{n}{n+1}$.

In fact, if there exists $1 \le t \le n$ with $\varepsilon_t = 1$, we choose $i = j = t$. Then

$$\left| \sum_{k=i}^{j}(\varepsilon_k - x_k) \right| = |\varepsilon_t - x_t| = \left| 1 - \frac{1}{n+1} \right| = \frac{n}{n+1}.$$

If $\varepsilon_t = 0$ for each $1 \le t \le n$, let $i = 1, j = n$. Then

$$\left| \sum_{k=i}^{j}(\varepsilon_k - x_k) \right| = \left| \sum_{k=1}^{n}(\varepsilon_k - x_k) \right| = n \left| 0 - \frac{1}{n+1} \right| = \frac{n}{n+1}.$$

Hence $\lambda \ge \dfrac{n}{n+1}$.

On the other hand, we show that if $\lambda = \dfrac{n}{n+1}$, then for any real numbers $x_1, x_2, \ldots, x_n \in [0, 1]$, there always exist $\varepsilon_1, \varepsilon_2, \ldots, \varepsilon_n \in \{0, 1\}$ with the desired property.

Let $S_0 = 0, S_1 = x_1, S_2 = x_1 + x_2, \ldots, S_n = x_1 + x_2 + \cdots + x_n$, and divide interval $[0, 1)$ into $n + 1$ smaller intervals $\left[0, \dfrac{1}{n+1} \right), \left[\dfrac{1}{n+1}, \dfrac{2}{n+1} \right), \ldots, \left[\dfrac{n}{n+1}, 1 \right)$. Then there must be an interval that does not contain any of $\{S_1\}, \{S_2\}, \ldots, \{S_n\}$ ($\{x\}$ denotes the fractional part of x). Let this interval be $\left[\alpha - \dfrac{1}{n+1}, \alpha \right)$.

Now we give the method of choosing $\varepsilon_1 \varepsilon_2, \ldots, \varepsilon_n$: for any $1 \le k \le n$, if there exists nonnegative integer m such that

$$S_{k-1} \le m + \alpha - \frac{1}{n+1} < m + \alpha \le S_k,$$

then let $\varepsilon_k = 1$, otherwise let $\varepsilon_k = 0$. We show that this selection meets the requirements, or in other words $|\sum_{k=i}^{j}(\varepsilon_k - x_k)| \le \dfrac{n}{n+1}$ for any $1 \le i < j \le n$.

In fact, by the definition of α, there exist nonnegative integers m, l such that

$$m - 1 + \alpha \le S_{i-1} \le m + \alpha - \frac{1}{n+1}, \quad l - 1 + \alpha \le S_j \le l + \alpha - \frac{1}{n+1}.$$

Hence

$$x_i + x_{i+1} + \cdots + x_j = S_j - S_{i-1} \in \left[l - m - \frac{n}{n+1}, l - m + \frac{n}{n+1} \right].$$

Then from the choice of $\varepsilon_1, \varepsilon_2, \ldots, \varepsilon_n$, we see that exactly $l - m$ among $\varepsilon_i, \ldots, \varepsilon_j$ take the value 1, so that $\varepsilon_i + \cdots + \varepsilon_j = l - m$.

Therefore,

$$(\varepsilon_i + \cdots + \varepsilon_j) - (x_i + \cdots + x_j) \in \left[-\frac{n}{n+1}, \frac{n}{n+1} \right].$$

Or equivalently

$$\left| \sum_{k=i}^{j} (\varepsilon_k - x_k) \right| \le \frac{n}{n+1}.$$

To summarize, the minimum of λ is $\dfrac{n}{n+1}$. $\qquad\square$

Remark The most difficult step is to view the left side of the inequality as the fractional part of the partial sum of x_k, and convert the inequality into a combinatorics problem. After this conversion, the difficulty of the problem is significantly decreased. The method used in this solution is similar to that of Problem B2 in the 67th Putnam Competition held in 2006:

Let $X = \{x_1, x_2, \ldots, x_n\}$ *be an n-element set of real numbers. Prove that there exists a nonempty subset S of X and an integer m, such that*

$$\left| m + \sum_{s \in S} s \right| \le \frac{1}{n+1}.$$

Second Day
(8:00 – 12:00; August 14, 2019)

5 As shown in Fig. 5.1, in an acute triangle ABC, $AB > AC$, and O, H are its circumcenter and orthocenter, respectively. The line passing through H and parallel to AB intersects AC at M, and the line passing through H and parallel to AC intersects AB at N. Let L be the reflection of H with respect to MN, and lines OL, AH intersect at K. Prove that K, M, L, N are concyclic.

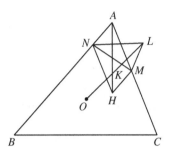

Fig. 5.1

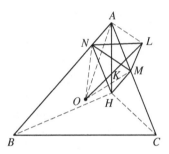

Fig. 5.2

Solution 1 As shown in Fig. 5.2, since L and H are symmetric with respect to HN, it follows that $ANHM$ is a parallelogram, so $\angle LNM = \angle HNM = \angle AMN$, $LN = HN = AM$, and consequently $\triangle AMN \cong \triangle LNM$. Hence $ANML$ is an isosceles trapezoid, $\angle NAM = \angle NLM$ and A, N, M, L are concyclic.

From the property of the orthocenter, we have $\angle HBN = \angle HCM$. Since $\angle BNH = \angle BAC = \angle CMH$, we have $\triangle BNH \sim \triangle CMH$, and $\dfrac{BN}{CM} = \dfrac{NH}{MH}$. Hence $BN \cdot NA = CM \cdot MA$, and N, M have the same power with respect to $\odot O$. In other words, $OM = ON$.

Let l be the perpendicular bisector of MN. Since $ANML$ is an isosceles trapezoid, points A, L are symmetric with respect to l, and N, M are symmetric with respect to l. Since O lies on l, we have $\angle KLM = \angle OAN = \angle KAM$, so that A, K, M, L are concyclic.

Therefore A, N, K, M, L all lie on the same circle, and K, M, L, N are concyclic. □

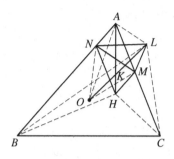

Fig. 5.3

Solution 2   As shown in Fig. 5.3, since L and H are symmetric with respect to HN, it follows that $ANHM$ is a parallelogram, so $\angle LNM = \angle HNM = \angle AMN$, $LN = HN = AM$, and consequently $\triangle AMN \cong \triangle LNM$. Hence $ANML$ is an isosceles trapezoid, $\angle NAM = \angle NLM$ and A, N, M, L are concyclic.

From the property of the orthocenter, we have $\angle HBN = \angle HCM$, and since $\angle BNH = \angle BAC = \angle CMH$, we have $\triangle BNH \sim \triangle CMH$, and $\dfrac{BN}{CM} = \dfrac{NH}{MH}$. Hence $\dfrac{LN}{LM} = \dfrac{HN}{HM} = \dfrac{BN}{CM}$.

Since $\angle ANL = \angle AML$ we have $\angle LNB = \angle LMC$, and $\triangle LNB \sim \triangle LMC$. Thus $\angle NBL = \angle MCL$, and L lies on $\odot O$, so that $OA = OL$.

Let l be the perpendicular bisector of AL. Since $ANML$ is an isosceles trapezoid, points N, M are symmetric with respect to l, and since O lies on l, we have $\angle KLM = \angle OAN = \angle KAM$. Hence A, K, M, L are concyclic.

Therefore A, N, K, M, L all lie on the same circle, and K, M, L, N are concyclic. $\qquad\square$

6   Suppose $a_1, a_2, \ldots, a_n$ are $n(n \geq 2)$ positive real numbers, such that $a_1 \leq a_2 \leq \cdots \leq a_n$. Prove that

$$\sum_{1 \leq i < j \leq n} (a_i + a_j)^2 \left(\frac{1}{i^2} + \frac{1}{j^2} \right) \geq 4(n - 1) \sum_{i=1}^{n} \frac{a_i^2}{i^2}.$$

(Contributed by Wang Guangting)

Solution   We use the method of mathematical induction to prove this inequality.

When $n = 2$, the desired inequality is

$$(a_1 + a_2)^2 \left(1 + \frac{1}{4}\right) \geq 4\left(a_1^2 + \frac{a_2^2}{4}\right)$$

$$\Leftrightarrow \frac{5}{4}(a_1 + a_2)^2 \geq 4a_1^2 + a_2^2$$

$$\Leftrightarrow 5(a_1 + a_2)^2 \geq 16a_1^2 + 4a_2^2$$

$$\Leftrightarrow a_2^2 + 10a_1a_2 \geq 11a_1^2.$$

Since $a_2 \geq a_1$, $a_2^2 \geq a_1^2$, $a_1a_2 \geq a_1^2$, the above inequality holds and the conclusion is proven for $n = 2$.

Suppose the inequality holds for n, then

$$\sum_{1 \leq i < j \leq n} (a_i + a_j)^2 \left(\frac{1}{i^2} + \frac{1}{j^2}\right) \geq 4(n-1) \sum_{i=1}^{n} \frac{a_i^2}{i^2}.$$

For $n + 1$, it suffices to prove that

$$\sum_{1 \leq i < j \leq n+1} (a_i + a_j)^2 \left(\frac{1}{i^2} + \frac{1}{j^2}\right) - \sum_{1 \leq i < j \leq n} (a_i + a_j)^2 \left(\frac{1}{i^2} + \frac{1}{j^2}\right)$$

$$\geq 4n \sum_{i=1}^{n+1} \frac{a_i^2}{i^2} - 4(n-1) \sum_{i=1}^{n} \frac{a_i^2}{i^2}.$$

Or equivalently,

$$\sum_{i=1}^{n} (a_i + a_{n+1})^2 \left(\frac{1}{i^2} + \frac{1}{(n+1)^2}\right) \geq 4 \sum_{i=1}^{n} \frac{a_i^2}{i^2} + \frac{4na_{n+1}^2}{(n+1)^2}. \qquad ①$$

Note that for $1 \leq i \leq n$ we have $a_i a_{n+1} \geq a_i^2$, so $(a_i + a_{n+1})^2 = a_i^2 + a_{n+1}^2 + 2a_i a_{n+1} \geq 3a_i^2 + a_{n+1}^2$. Hence

$$\text{LHS of (1)} \geq \sum_{i=1}^{n} (3a_i^2 + a_{n+1}^2) \left(\frac{1}{i^2} + \frac{1}{(n+1)^2}\right)$$

$$= 3 \sum_{i=1}^{n} \left(\frac{1}{i^2} + \frac{1}{(n+1)^2}\right) a_i^2 + \left(\sum_{i=1}^{n} \frac{1}{i^2} + \frac{1}{(n+1)^2}\right) a_{n+1}^2.$$

Then it suffices to show that

$$3\sum_{i=1}^{n}\left(\frac{1}{i^2}+\frac{1}{(n+1)^2}\right)a_i^2+\left(\sum_{i=1}^{n}\frac{1}{i^2}+\frac{1}{(n+1)^2}\right)a_{n+1}^2$$

$$\geq 4\sum_{i=1}^{n}\frac{a_i^2}{i^2}+\frac{4na_{n+1}^2}{(n+1)^2}\Leftrightarrow\sum_{i=1}^{n}\left(\frac{3}{(n+1)^2}-\frac{1}{i^2}\right)a_i^2$$

$$+\left(\sum_{i=1}^{n}\frac{1}{i^2}-\frac{3n}{(n+1)^2}\right)a_{n+1}^2\geq 0.\qquad\qquad(2)$$

Note that $\sum_{i=1}^{n}\frac{1}{i^2}-\frac{3n}{(n+1)^2}\geq 1-\frac{3n}{(n+1)^2}>0$, $\frac{3}{(n+1)^2}-\frac{1}{i^2}$ is monotonically increasing with i, and a_i^2 is also monotonically increasing with i, so Chebyshev's inequality gives

$$\sum_{i=1}^{n}\left(\frac{3}{(n+1)^2}-\frac{1}{i^2}\right)a_i^2\geq\frac{1}{n}\left[\sum_{i=1}^{n}\left(\frac{3}{(n+1)^2}-\frac{1}{i^2}\right)\right]\cdot\left(\sum_{i=1}^{n}a_i^2\right)$$

$$=-\left(\sum_{i=1}^{n}\frac{1}{i^2}-\frac{3n}{(n+1)^2}\right)\cdot\left(\frac{1}{n}\sum_{i=1}^{n}a_i^2\right)$$

$$\geq-\left(\sum_{i=1}^{n}\frac{1}{i^2}-\frac{3n}{(n+1)^2}\right)a_{n+1}^2.$$

Hence ② holds, and the desired inequality is true for $n+1$.

Therefore, the induction is complete, and the original inequality is proven. □

Remark (by Zhou Tianyou) During the contest, some students gave a very concise proof of (2). Note that it is equivalent to

$$\sum_{i=1}^{n}\frac{a_{n+1}^2-a_i^2}{i^2}\geq\sum_{i=1}^{n}\frac{3(a_{n+1}^2-a_i^2)}{(n+1)^2}.$$

While

$$\sum_{i=1}^{n}\frac{a_{n+1}^2-a_i^2}{i^2}\geq a_{n+1}^2-a_1^2\geq\frac{3n(a_{n+1}^2-a_1^2)}{(n+1)^2}\geq\sum_{i=1}^{n}\frac{3(a_{n+1}^2-a_i^2)}{(n+1)^2}.$$

Therefore, the desired inequality holds.

7 Prove that for any positive integer k, there exist at most finitely many sets T with the following properties:

(1) T consists of finitely many prime numbers;

(2) $\prod\limits_{p \in T} p \,\Big|\, \prod\limits_{p \in T} (p + k)$.

(Contributed by Yang Mingliang)

Solution We prove by contradiction. Suppose there are infinitely many such Ts. Since a finite set only admits finitely many subsets, it follows that for any positive integer M, there is some set T with the desired properties such that T has an element greater than M (since otherwise any such T is a subset of $\{1, 2, \ldots, M\}$, a contradiction).

Choose $M = 2k^2 + 2k$. Then there exists set T with the desired properties, and containing an element greater than M. Let q be the greatest element of T.

Next we prove by induction that for $i = 0, 1, \ldots, k$, number $q - ik$ belongs to T. The assertion holds when $i = 0$. Suppose the assertion is true for i, consider $i + 1$. We have

$$(q - ik) \,\Big|\, \prod_{p \in T} (p + k) \,.$$

From the induction hypothesis, $q - ik$ is a prime number, so there is some $p \in T$ such that $p + k$ is divisible by $q - ik$. Since $q > M$, we have

$$2(q - ik) - (q + k) = q - (2i + 1)k > 2k^2 + 2k - (2k + 1)k > 0.$$

Hence $p + k \le q + k < 2(q - ik)$, and the only possibility is $q - ik = p + k$, so that $p = q - (i + 1)k$ is an element of T. Therefore the assertion holds for $i + 1$, and the induction is complete.

Thus $q, q - k, q - 2k, \ldots, q - k^2$ are all elements of T. Note that $\gcd(k, k + 1) = 1$, so $q, q - k, q - 2k, \ldots, q - k^2$ compose a complete residue system modulo $k + 1$, and there exists $0 \le i \le k$ such that $q - ik$ is a multiple of $k + 1$. However, since $q - ik \ge 2k^2 + 2k - k^2 = k^2 + 2k > k + 1$, and $k + 1 > 1$, we see that $q - ik$ is composite, leading to a contradiction.

In summary, there are finitely many such Ts. $\square$

8 We call a set that can be presented in the form $\{x, 2x, 3x\}$ a "good" set (where x is a nonzero real number). For a given integer $n(n \geq 3)$, what is the maximum number of "good" subsets that an n-element set of positive integers may have?

(Contributed by Yang Mingliang)

Solution 1 The desired maximum is $n - \left\lceil \dfrac{\sqrt{8n+1} - 1}{2} \right\rceil$. (Here $\lceil y \rceil$ means the smallest integer that is not smaller than real number y.)

We first show that for any n-element set S of positive integers, the number of "good" subsets of S cannot exceed $n - \left\lceil \dfrac{\sqrt{8n+1} - 1}{2} \right\rceil$.

Let $t = \left\lceil \dfrac{\sqrt{8n+1} - 1}{2} \right\rceil$ $(t \in \mathbb{N}^*)$, then $\dfrac{\sqrt{8n+1} - 1}{2} \leq t < \dfrac{\sqrt{8n+1} + 1}{2}$, which is equivalent to

$$(2t - 1)^2 < 8n + 1 \leq (2t + 1)^2$$

$$\Leftrightarrow \frac{1}{2}t(t-1) < n \leq \frac{1}{2}t(t+1).$$

Thus we may assume $n = \dfrac{1}{2}t(t-1) + r$, where $1 \leq r \leq t(r \in \mathbb{N}^*)$.

Next we define an equivalence relation "$\sim$" as follows: the numbers x, y satisfy $x \sim y$ if and only if there exists $\alpha \in \mathbb{Z}$ such that $x = 2^\alpha y$. For a positive odd number m, denote $S_m = \{x \mid x \in S, x \sim m\}$. Thus each element of S belongs to a unique S_m. Since S is finite, it follows that only finitely many S_ms are nonempty. Let them be $m_1 < m_2 < \cdots < m_k$, so $S = S_{m_1} \cup S_{m_2} \cup \cdots \cup S_{m_k}$.

Let $T_{m_i} = \{x \mid x \in S_{m_i}, 2x \in S, 3x \in S\}(1 \leq i \leq k)$. For any "good" subset of S, namely $\{y, 2y, 3y\}$, y must be an element of $T_{m_1} \cup T_{m_2} \cup \cdots \cup T_{m_k}$, and they are in one-to-one correspondence. Thus S has exactly $|T_{m_1}| + |T_{m_2}| + \cdots + |T_{m_k}|$ "good" subsets.

Then we show that $\sum\limits_{i=1}^{k} |T_{m_i}| \leq n - t$. Note that for $x \in T_{m_i}$, we have $2x \in S$, so $2x \in S_{m_i}$. Let x_0 be the element in S_{m_i} with the highest power of 2. Obviously, $x_0 \notin T_{m_i}$. Therefore,

$$|T_{m_i}| \leq |S_{m_i}| - 1. \qquad \qquad (1)$$

Further, for $x \in T_{m_i}$, we have $3x \in S$, so

$$|S_{3m_i}| \geq |T_{m_i}|. \qquad (2)$$

Hence, we have the followings:

(1) If $k \geq t$, then

$$\sum_{i=1}^{k} |T_{m_i}| \leq \sum_{i=1}^{k} (|S_{m_i}| - 1) = |S| - k = n - k \leq n - t$$

(2) If $k \leq t - 1$, and $|T_{m_i}| \leq t - i - 1$ for $1 \leq i \leq k$, then

$$\sum_{i=1}^{k} |T_{m_i}| \leq \sum_{i=1}^{k} (t - i - 1) \leq \frac{1}{2}(t-1)(t-2) < n - t$$

(3) If $k \leq t - 1$, and $|T_{mj}| \geq t - j$ for some $1 \leq j \leq k$. Note that

$$S \supseteq (S_{m_1} \cup S_{m_2} \cup \cdots \cup S_{m_j} \cup S_{3m_j} \cup S_{3m_{j+1}} \cup \cdots \cup S_{3m_k})$$

so we have

$$n \geq \sum_{i=1}^{j} |S_{m_i}| + \sum_{i=j}^{k} |S_{3m_i}|$$

$$\geq \sum_{i=1}^{j} |T_{m_i}| + j + (t - j) + \sum_{i=j+1}^{k} |T_{m_i}|$$

$$= \sum_{i=1}^{k} |T_{m_i}| + t.$$

Therefore,

$$\sum_{i=1}^{k} |T_{m_i}| \leq n - t.$$

Combining (1), (2), (3), we conclude that the number of "good" subsets of S is not greater than $n - t$.

Finally we give a construction of P with exactly $n - t$ "good" subsets. Let P be an n-element set of positive integers which is the union of the following t sets $P_1, P_2, \ldots, P_t$: if $1 \leq i \leq r$, then $P_i = \{2^i 3^\beta \mid 1 \leq \beta \leq t + 1 - i\}$; if $r + 1 \leq i \leq t$, then $P_i = \{2^i 3^\beta \mid 1 \leq \beta \leq t - i\}$.

Then $|P| = \frac{1}{2}t(t-1)+r = n$, and P has subsets $\{2^{\alpha}3^{\beta}, 2^{\alpha+1}3^{\beta}, 2^{\alpha}3^{\beta+1}\}$ where $1 \le \alpha \le r-1, 1 \le \beta \le t-\alpha$ or $r \le \alpha \le t-1, 1 \le \beta \le t-\alpha-1$. The number of such $\{\alpha, \beta\}$ is

$$(0 + 1 + \cdots + t - 2) + (r-1) = \frac{1}{2}(t-1)(t-2) + r - 1 = n - t.$$

Therefore, the desired maximum is $n - \left\lceil \dfrac{\sqrt{8n+1}-1}{2} \right\rceil$.

Solution 2 The desired maximum is $n - \left\lceil \dfrac{\sqrt{8n+1}-1}{2} \right\rceil$.

The construction of the example is the same as in Solution 1.

Next we will show that the number of "good" subsets is at most $n - \left\lceil \dfrac{\sqrt{8n+1}-1}{2} \right\rceil$. Let the n numbers correspond to n vertices in a graph. If x, y satisfy $2x = y$, then we draw a red arrow from y to x; if $3x = y$, then we draw a blue arrow from y to x. Thus each "good" subset corresponds to a shape consisting of a red arrow and a blue arrow starting at the same vertex.

It is easily seen that a red arrow and a blue arrow cannot have the same endpoints. The red arrows compose several disjoint chains, and the blue arrows also compose several disjoint chains (where we regard an isolated vertex as a single chain). We have

$$|\text{red arrows}| = n - |\text{red chains}|, \quad |\text{blue arrows}| = n - |\text{blue chains}|.$$

Suppose there are a vertices that do not shoot out a red arrow and a blue arrow simultaneously, so that the remaining $n - a$ vertices shoot out both a red arrow and a blue arrow. For each of these $n - a$ vertices we do the following operation: while the vertex we arrive still shoots out a red arrow, then we go through the red arrow to reach the next vertex. Then we will finally reach some vertex that does not shoot out a red arrow. Similarly we can reach a vertex that does not shoot out a blue arrow if we go through blue arrows. Thus the initial vertex is mapped to an unordered pair of terminal vertices. Then we show that this mapping is an injection.

Suppose the number x stops at u when going through red arrows, and stops at v when going through blue arrows. Then we write x, u, v in the form of $2^{\alpha}3^{\beta}m$ (where α, β are nonnegative integers, and m is coprime to 6). Let $x = 2^{\alpha_1}3^{\beta_1}m_1$, $u = 2^{\alpha_2}3^{\beta_2}m_2, v = 2^{\alpha_3}3^{\beta_3}m_3$. Then necessarily, $m_1 = m_2 = m_3, \alpha_1 = \min\{\alpha_2, \alpha_3\}, \beta_1 = \min\{\beta_2, \beta_3\}$. Hence x is uniquely determined by u, v, and the mapping above is an injection.

Therefore, $n - a \leq \dfrac{a(a-1)}{2}$, so that $\dfrac{a(a+1)}{2} \geq n$, which implies

$$a \geq \left\lceil \frac{\sqrt{8n+1}-1}{2} \right\rceil.$$

And the number of "good" subsets is at most $n - a \leq n - \left\lceil \dfrac{\sqrt{8n+1}-1}{2} \right\rceil$.

□

China Southeastern
Mathematical Olympiad

2018 (Jinjiang, Fujian)

From July 28 to August 2, 2018, the 15th China Southeastern Mathematical Olympiad (CSMO) was held at the Yangzheng Middle School in Jinjiang City, Fujian Province, China. 1336 players, coming from various provinces and cities in the Mainland China, Hong Kong and Macao, and some countries in Southeast Asia, participated in the competition.

The main examination committee members of this competition are:

Wang Changping (Fujian Normal University);

Zhang Pengcheng (Fujian Normal University);

Chang An (Fuzhou University);

Xiong Bin (East China Normal University);

He Yijie (East China Normal University);

Lin Tianqi (East China Normal University);

Li Shenghong (Zhejiang University);

Yang Xiaoming (Zhejiang University);

Liu Bin (Peking University);

Hu Zhiming (Tsinghua University);

Tao Pingsheng (Jiangxi Science and Technology Normal University).

The competition is divided into two levels: the first grade and the second grade of senior high school.

10th Grade
First Day
(8:00 – 12:00; July 30, 2018)

1 Let c be a real number. If there exists $x \in [1,2]$ such that $\max\left\{\left|x + \dfrac{c}{x}\right|, \left|x + \dfrac{c}{x} + 2\right|\right\} \geq 5$, find the value range of c. (Here $\max\{a, b\}$ denotes the greater number of real numbers a and b.) (Contributed by Yang Xiaoming)

Solution 1   Note that

$$\max\left\{\left|x + \frac{c}{x}\right|, \left|x + \frac{c}{x} + 2\right|\right\} = \max\left\{\left|x + \frac{c}{x} + 1 - 1\right|, \left|x + \frac{c}{x} + 1 + 1\right|\right\}$$

$$= \left|x + \frac{c}{x} + 1\right| + 1.$$

So the condition reduces to that there exists $x \in [1,2]$ such that $\left|x + \dfrac{c}{x} + 1\right| \geq 4$.

Let $g(x) = x + \dfrac{c}{x} + 1$, where $x \in [1,2]$. We discuss based on the following cases.

Case 1: $c = 0$. Then $g(x) = x + 1$, and for $x \in [1,2]$ we have $|g(x)| \leq 3 < 4$, which is contradictory to the given condition.

Case 2: $c > 0$. Then $g(x) > 0$, and it follows from the graph of $g(x)$ that

$$|g(x)|_{\max} = \max\left\{2 + c, 3 + \frac{c}{2}\right\}.$$

Thus either $2 + c \geq 4$ or $3 + \dfrac{c}{2} \geq 4$, and the resulting range is $c \geq 2$.

Case 3: $c < 0$. Then $g(x)$ is monotonically increasing, and in this case

$$|g(x)|_{\max} = \max\left\{|2 + c|, \left|3 + \frac{c}{2}\right|\right\}.$$

Since $c < 0$, we have $2 + c \leq -4$ or $3 + \frac{c}{2} \leq -4$, which gives $c \leq -6$.

In summary, the value range of c is $(-\infty, -6] \cup [2, +\infty)$.

Solution 2   The converse of the assertion is that for all $x \in [1,2]$ we have $\max\left\{\left|x + \dfrac{c}{x}\right|, \left|x + \dfrac{c}{x} + 2\right|\right\} < 5$, which is equivalent to saying that $\left|x + \dfrac{c}{x}\right| < 5$ and $\left|x + \dfrac{c}{x} + 2\right| < 5$. Solving the inequalities we obtain $-5 < x + \dfrac{c}{x} < 3$. Since $x > 0$, we have $-x(5 + x) < c < x(3 - x)$.

Then since this inequality holds for all $x \in [1, 2]$, it follows that

$$[-x(5+x)]_{\max} < c < [x(3-x)]_{\min}.$$

This implies that $c \in (-6, 2)$. Therefore, the desired value range of c is

$$\mathbb{R}/(-6, 2) = (-\infty, -6] \cup [2, +\infty).$$ □

2 In the coordinate plane, we call a point a rational point if both of its coordinates are rational numbers, and otherwise we call it an irrational point. Suppose there is a regular pentagon in the plane. Then among its vertices, are there more rational points than irrational points or more irrational points than rational points? Prove your result. (Contributed by Wang Changping)

Solution We shall prove that for any regular pentagon in the coordinate plane, there are always more irrational points than rational points among its vertices.

In fact, let the side length and diagonal length of the pentagon be a, b, respectively. Then we have $\dfrac{a}{b} = \dfrac{\sqrt{5}-1}{2}$, so

$$\frac{a^2}{b^2} = \left(\frac{a}{b}\right)^2 = \left(\frac{\sqrt{5}-1}{2}\right)^2 = \frac{3-\sqrt{5}}{2},$$

which is irrational. Suppose there are three rational points in the vertices of the pentagon, then the pairwise distances of these vertices must contain both a and b. However, the square of the distance between any two rational points is a rational number, so both a^2 and b^2 are rational, which is contradictory to the fact that $\dfrac{a^2}{b^2}$ is irrational. Therefore, a regular pentagon in the coordinate plane cannot have more than 2 rational vertices, so it has more irrational vertices than rational vertices. □

3 As shown in Fig. 3.1, acute triangle ABC is inscribed to $\odot O\,(AB < AC)$. The bisector of $\angle BAC$ intersects BC at T, and M is the midpoint of AT. Point P lies inside the triangle, such that $PB \perp PC$. Construct the line passing through P and perpendicular to AP, and let D, E be points on this line (which are different from P) such that $BD = BP, CE = CP$. Suppose line AO bisects segment DE. Prove that line AO is tangent to the circumcircle of $\triangle AMP$.

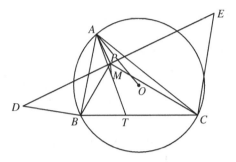

Fig. 3.1

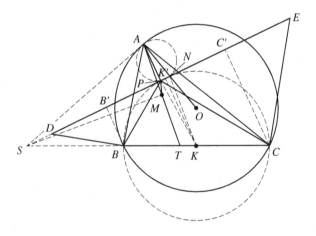

Fig. 3.2

Solution   Let N be the intersection of AO and DE, which is the midpoint of DE. We first show that B, P, N, C are concyclic. Let K be the midpoint of BC, and let B', C', K' be the projections of B, C, K on the line DE. Then they are the midpoints of $PD, PE, B'C'$, respectively.

Thus,

$$PB' = \frac{1}{2}DP = \frac{1}{2}DE - \frac{1}{2}PE = NE - C'E = NC'.$$

This means that K' is the midpoint of PN, i.e., $KP = KN$. From $PB \perp PC$ we see that $KB = KP = KC$, and P, N both lie on the circle with diameter BC.

Apparently, P also lies on the circle with diameter AN, which is tangent to $\odot O$. We note that the tangent line to $\odot O$ at A, the line PN and the line BC are the three pairwise radical axes of the circles $\odot O, \odot K, \odot AN$, so they are concurrent at the radical center S. By the property of the chord-tangent angle we have

$$\angle SAT = \angle SAB + \angle BAT = \angle ACB + \angle CAT = \angle ATS,$$

so $SA = ST$. Since M is the midpoint of AT, we have $\angle AMS = 90°$. Combining with $\angle APS = 90°$, we obtain that A, P, M, S are concyclic. Hence,

$$\angle AMP = \angle ASP = 90° - \angle SAP = \angle OAP.$$

Therefore AO is tangent to the circumcircle of $\triangle AMP$. $\qquad \square$

4 Determine whether exists set $A \subseteq \mathbb{N}^*$, such that for any positive integer n, $A \cap \{n, 2n, 3n, \ldots, 15n\}$ has exactly one element. Prove your result. (Contributed by He Yijie)

Solution For any positive integer l, if $l = 2^a 3^b 5^c 7^d 11^e 13^f t$ (where a, b, c, d, e, f are nonnegative integers and t is a positive integer that is coprime to $2, 3, 5, 7, 11, 13$), then define

$$f(l) = a + 4b + 9c + 11d + 7e + 14f.$$

For any positive integer n, we will show that $f(kn)$ $(k = 1, 2, \ldots, 15)$ compose a complete residue system modulo 15. In fact, let $n = 2^a 3^b 5^c 7^d 11^e 13^f t$. Then for $k = 1, 2, \ldots, 15$, we denote $\triangle_k = f(kn) - f(n)$, and let the factorization of k be $k = 2^{\alpha_1} 3^{\alpha_2} 5^{\alpha_3} 7^{\alpha_4} 11^{\alpha_5} 13^{\alpha_6}$. Then

$$\triangle_k = f(2^{a+\alpha_1} 3^{b+\alpha_2} 5^{c+\alpha_3} 7^{d+\alpha_4} 11^{e+\alpha_5} 13^{f+\alpha_6} t) - f(2^a 3^b 5^c 7^d 11^e 13^f t)$$

$$= \alpha_1 + 4\alpha_2 + 9\alpha_3 + 11\alpha_4 + 7\alpha_5 + 14\alpha_6.$$

Thus we compute that

$$\triangle_1 = 0, \ \triangle_2 = 1, \ \triangle_3 = 4, \ \triangle_4 = 2, \ \triangle_5 = 9, \ \triangle_6 = 5, \ \triangle_7 = 11, \ \triangle_8 = 3,$$

$$\triangle_9 = 8, \ \triangle_{10} = 10, \ \triangle_{11} = 7, \ \triangle_{12} = 6, \ \triangle_{13} = 14, \ \triangle_{14} = 12, \ \triangle_{15} = 13.$$

Hence, $f(kn)$ $(k = 1, 2, \ldots, 15)$ compose a complete residue system modulo 15.

Further, let

$$A = \{n \in \mathbb{N}^* \,|\, f(n) \equiv 0 \,(\mathrm{mod}\,15)\}.$$

Then it follows that for any positive integer n, $A \cap \{n, 2n, \ldots, 15n\}$ has exactly one element. $\qquad \square$

Second Day
(8:00 – 12:00; July 31, 2018)

5 Let $\{a_n\}$ be a sequence of nonnegative real numbers. Define

$$X_k = \sum_{i=1}^{2^k} a_i, \quad Y_k = \sum_{i=1}^{2^k} \left[\frac{2^k}{i}\right] a_i, \quad k = 0, 1, 2, \ldots,$$

where $[\alpha]$ denotes the greatest integer that does not exceed real number α. Prove that for any positive integer n,

$$X_n \leq Y_n - \sum_{i=0}^{n-1} Y_i \leq \sum_{i=0}^{n} X_i.$$

(Contributed by Li Shenghong, He Yijie)

Solution For any positive integer k, we have

$$Y_k - 2Y_{k-1} = \left(\sum_{i=1}^{2^{k-1}} \left[\frac{2^k}{i}\right] a_i + \sum_{i=2^{k-1}+1}^{2^k} \left[\frac{2^k}{i}\right] a_i\right) - \sum_{i=1}^{2^{k-1}} 2\left[\frac{2^{k-1}}{i}\right] a_i$$

$$= \sum_{i=1}^{2^{k-1}} \left(\left[2 \cdot \frac{2^{k-1}}{i}\right] - 2\left[\frac{2^{k-1}}{i}\right]\right) a_i + \sum_{i=2^{k-1}+1}^{2^k} a_i. \qquad \textcircled{1}$$

Since $a_i \geq 0 (1 \leq i \leq 2^{k-1})$, combining with $[2\alpha] - 2[\alpha] = [2\{\alpha\}] \in \{0, 1\}$ $(\alpha \in \mathbb{R})$, we obtain that

$$0 \leq \sum_{i=1}^{2^{k-1}} \left(\left[2 \cdot \frac{2^{k-1}}{i}\right] - 2\left[\frac{2^{k-1}}{i}\right]\right) a_i \leq \sum_{i=1}^{2^{k-1}} a_i = X_{k-1}.$$

Also since $\sum_{i=2^{k-1}+1}^{2^k} a_i = X_k - X_{k-1}$, it follows from $\textcircled{1}$ that

$$X_k - X_{k-1} \leq Y_k - 2Y_{k-1} \leq X_k - X_{k-1} + X_{k-1} = X_k.$$

Further, we have

$$X_n - X_0 = \sum_{i=1}^{n} (X_i - X_{i-1}) \leq \sum_{i=1}^{n} (Y_i - 2Y_{i-1}) \leq \sum_{i=1}^{n} X_i. \qquad \textcircled{2}$$

Note that $Y_0 = X_0 = a_1$, so

$$\sum_{i=1}^{n} (Y_i - 2Y_{i-1}) = Y_n - \left(\sum_{i=1}^{n-1} Y_i\right) - 2Y_0 = Y_n - \left(\sum_{i=0}^{n-1} Y_i\right) - X_0.$$

Combining with ②, we conclude that

$$X_n \le Y_n - \sum_{i=0}^{n-1} Y_i \le X_0 + \sum_{i=1}^{n} X_i = \sum_{i=0}^{n} X_i.$$

Thus the desired inequality is proven. □

⑥ As shown in Fig. 6.1, in $\triangle ABC$, $AB = AC$. The center of $\odot O$ is the midpoint of BC. AB, AC are tangent to the circle at E, F, respectively. Point G lies on $\odot O$, such that $AG \perp EG$. The tangent line to $\odot O$ at G intersects AC at K. Prove that line BK bisects segment EF. (Contributed by Zhang Pengcheng)

Solution As shown in Fig. 6.2, let D be the intersection of EF and BK. Extend KG to intersect AB at M. Since $ME = MG$ and $AG \perp EG$, M is the midpoint of AE. We apply Menelaus' theorem to triangle ABK and line EDF, and obtain

$$\frac{AE}{EB} \cdot \frac{BD}{DK} \cdot \frac{KF}{FA} = 1.$$

Since $AE = AF$, it follows that

$$\frac{BD}{BK} = \frac{BE}{FK}. \qquad ①$$

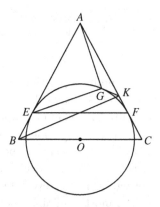

Fig. 6.1

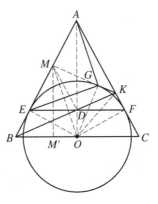

Fig. 6.2

Let M' be the projection of M on line BC. We have

$$\angle BOM = \angle BOE + \angle MOE = \angle BAO + \frac{1}{2}\angle EOG$$

$$= \frac{1}{2}\angle BAC + \frac{1}{2}\angle AMK = \frac{1}{2}\angle MKC = \angle CKO,$$

as well as $\angle B = \angle C$, so $\triangle BOM \sim \triangle CKO$, and MM' and OF are the corresponding altitudes. Hence,

$$\frac{BE}{FK} = \frac{CF}{FK} = \frac{BM'}{M'O} = \frac{BM}{MA}. \qquad \textcircled{2}$$

Combining $\textcircled{1}$, $\textcircled{2}$ we obtain $\dfrac{BD}{DK} = \dfrac{BM}{MA}$, which means $MD \parallel AK$. Also M is the midpoint of AE, so MD is the median line of $\triangle AEF$, and D is the midpoint of EF.   □

7 There are 24 people attending a meeting. Each pair of them either shake hands once with each other, or do not. After the meeting, we find that there are totally 216 pairs of people having shaken hands with each other, and for each pair among them, saying the pair of P and Q, at most 10 out of the remaining 22 people have shaken hands with exactly one of P, Q. A "friend circle" refers to a set of three people from the meeting, such that every two of them have shaken hands with each other. Find the minimum of the number of "friend circles" among these 24 people. (Contributed by Chang An)

Solution 1 Let the 24 people be $v_1, v_2, \ldots, v_{24}$, and v_i has shaken hands with d_i people for $i = 1, 2, \ldots, 24$. Define the set

$$E = \{\{v_i, v_j\} \mid v_i \text{ has shaken hands with } v_j\}$$

For each $e = \{v_i, v_j\} \in E$, let $t(e)$ be the number of "friend circles" containing e. Then it follows from the condition that

$$(d_i - 1) + (d_j - 1) \leq 2t(e) + 10.$$

Thus

$$\sum_{e = \{v_i, v_j\} \in E} (d_i + d_j) \leq \sum_{e \in E} (2t(e) + 12). \qquad \text{①}$$

Note that in expression $\sum_{e = \{v_i, v_j\} \in E} (d_i + d_j)$, number d_i is counted d_i times, so the left side of ① equals $\sum_{i=1}^{24} d_i^2$. Let T be the number of friend circles, then the right side of ① equals $6T + 12 \times 216$. Hence

$$\sum_{i=1}^{24} d_i^2 \leq 6T + 12 \times 216.$$

Since $\sum_{i=1}^{24} d_i^2 \geq \frac{1}{24} \left(\sum_{i=1}^{24} d_i \right)^2 = \frac{(2 \times 216)^2}{24}$, we also have

$$T \geq \frac{1}{6} \left[\frac{(2 \times 216)^2}{24} - 12 \times 216 \right] = 864.$$

On the other hand, we divide the 24 people into 4 groups, with 6 people in each group, and stipulate that a pair of people shake hands if and only if they belong to different groups. Then the total number of hand shakings is $\frac{1}{2} \times 24 \times 18 = 216$. Further, for each pair P, Q who have shaken hands, those who have shaken hands with exactly one of them are exactly the remaining 10 people in the two groups containing P and Q. In this case, the total number of friend circles is $\binom{4}{3} \times 6^3 = 864$.

In summary, the minimal number of friend circles among these people is 864. □

Solution 2 Let the 24 people correspond to 24 vertices $A_1, A_2, \ldots, A_{24}$. If two people have shaken hands with each other, then we connect an edge between the corresponding vertices. Thus we have constructed a simple graph. If A_i is connected to A_j, A_k with edges, then we call $\angle A_j A_i A_k$ and "angle" (note that $\angle A_j A_i A_k$ and $\angle A_k A_i A_j$ refer to the same angle).

Further, if A_j, A_k are also connected by an edge, we will call $\angle A_j A_i A_k$ a "first-type angle", and otherwise a "second-type angle".

Let A_i have degree d_i in the graph ($1 \leq i \leq 24$), and the total number of first-type angles and second-type angles are m_1, m_2, respectively. Also let T be the number of friend circles. If $A_i A_j$ is an edge, then the number of second-type angles containing $A_i A_j$ as one of its sides does not exceed 10. Hence,

$$m_2 \leq \frac{1}{2} \times 10 \times 216 = 1080.$$

The total number of angles is

$$
\begin{aligned}
m_1 + m_2 &= \sum_{i=1}^{24} C_{d_i}^2 = \frac{1}{2}\left[\sum_{i=1}^{24} d_i^2 - \left(\sum_{i=1}^{24} d_i\right)\right] \\
&\geq \frac{1}{2}\left[\frac{1}{24}\left(\sum_{i=1}^{24} d_i\right)^2 - \left(\sum_{i=1}^{24} d_i\right)\right] \\
&= \frac{1}{2}\left(\sum_{i=1}^{24} d_i\right) \cdot \left[\frac{1}{24}\left(\sum_{i=1}^{24} d_i\right) - 1\right] \\
&= 216 \times \left(\frac{1}{24} \times 2 \times 216 - 1\right) = 216 \times 17.
\end{aligned}
$$

Since $m_1 = 3T$, it follows that

$$T = \frac{1}{3}m_1 \geq \frac{1}{3} \times (216 \times 17 - 1080) = 864.$$

The remaining part is the same as in Solution 1.   □

8 Let m be a given positive integer. For positive integer l, we write

$$A_l = (4l + 1) \times (4l + 2) \times \cdots \times (4(5^m + 1)l).$$

Prove that there exist infinitely many positive integers l such that $5^{5^m l} \mid A_l$, $5^{5^m l + 1} \nmid A_l$. Also find the minimum of such l. (Contributed by Yang Xiaoming)

Solution For positive integers $p \geq 2$ and $n \geq 1$, we denote the sum of digits in the base-p representation of n by $S_p(n)$. If p is a prime number, we denote the power of p in the standard factorization of n by $v_p(n)$.

Lemma *For positive integer n and prime number p,*

$$v_p(n!) = \sum_{i=1}^{\infty} \left[\frac{n}{p^i}\right] = \frac{n - S_p(n)}{p - 1}.$$

Proof of lemma For any positive integer i, let U_i be the set of multiples of p^i in $1, 2, \ldots, n$. Then for $u \leq n$, $v_p(n)$ equals the number of U_i that contain u. Hence,

$$v_p(n!) = \sum_{m=1}^{n} v_p(m) = \sum_{i=1}^{\infty} |U_i| = \sum_{i=1}^{\infty} \left[\frac{n}{p^i}\right].$$

Let $n = (a_n a_{n-1} \cdots a_0)_p$ be the base-p representation of n, where $0 \leq a_i \leq p - 1$ $(0 \leq i \leq n)$. Then

$$v_p(n!) = \sum_{i=1}^{\infty} \left[\frac{n}{p^i}\right] = (a_n a_{n-1} \cdots a_1)_p + (a_n a_{n-1} \cdots a_2)_p + \cdots + (a_n)_p$$

$$= a_1 + (1 + p)a_2 + (1 + p + \cdots + p^{n-1})a_n$$

$$= \frac{(p - 1)a_1 + (p^2 - 1)a_2 + \cdots + (p^n - 1)a_n}{p - 1}$$

$$= \frac{(a_0 + a_1 p + \cdots + a_n p^n) - (a_0 + a_1 + \cdots + a_n)}{p - 1}$$

$$= \frac{n - S_p(n)}{p - 1}.$$

Thus, the lemma is proven.

Now come back to the original problem, where the given condition implies $v_5(A_l) = 5^m l$. From the lemma, we see that

$$v_5(A_l) = v_5\left[\frac{(4(5^m + 1)l)!}{(4l)!}\right] = v_5[(4(5^m + 1)l)!] - v_5[(4l)!]$$

$$= \frac{4(5^m + 1)l - S_5(4(5^m + 1)l)}{5 - 1} - \frac{4l - S_5(4l)}{5 - 1}$$

$$= 5^m \cdot l - \frac{S_5(4(5^m + 1)l) - S_5(4l)}{4}.$$

Let $n = 4l$. Then in order to satisfy the given condition, we should have

$$S_5((5^m + 1)n) = S_5(n). \qquad \qquad \textcircled{1}$$

If n has fewer than $m + 1$ digits in its base-5 representation, then

$$S_5((5^m + 1)n) = S_5(5^m n + n) = S_5(5^m n) + S_5(n) = 2S_5(n) > S_5(n),$$

which is a contradiction.

If n has exactly $m + 1$ digits in its base-5 representation, we show that only $n = 5^{m+1} - 1$ satisfies ①.

Let $n = (a_m a_{m-1} \cdots a_0)_5$, where $0 \le a_i \le 4 \, (i = 0, 1, \ldots, m)$ and $a_m \ne 0$. Suppose the addition $5^m n + n$ is carried t times when calculated in base 5. Since n has $m + 1$ digits in base 5, we have $t \le m + 1$. Now we have

$$S_5((5^m + 1)n) = S_5(5^m n + n) = S_5(5^m n) + S_5(n) - 4t$$

$$= 2S_5(n) - 4t = S_5(n).$$

So $S_5(n) = 4t$. This implies that the carry-over in addition must happen, so $a_0 + a_m \ge 5$, and $a_1 = a_2 = \cdots = a_{t-1} = 4$.

If $t < m + 1$, then

$$S_5(n) \ge (a_0 + a_m) + (a_1 + a_2 + \cdots + a_{t-1}) \ge 5 + 4(t - 1) > 4t,$$

which is a contradiction. Hence, $t = m + 1$ and $a_1 = a_2 = \cdots = a_m = 4$. In this case, since $a_0 + a_m \ge 5$ we have $a_0 > 0$, and combining with $n = 4l$ we obtain that $a_0 = 4$, so $n = (\underbrace{44 \cdots 4}_{m+1 \text{ copies}})_5 = 5^{m+1} - 1$, and we can check that ① holds. Therefore, $l = \dfrac{5^{m+1} - 1}{4}$ is the minimal integer that satisfies our condition, and we denote this number as l_0.

Further, for every $\alpha \in \mathbb{N}$, we also have

$$S_5((5^m + 1)4 \cdot 5^\alpha l_0) = S_5((5^m + 1)4l_0) = S_5(4l_0) = S_5 \left(4 \cdot 5^\alpha l_0 \right).$$

So $l = 5^\alpha l_0$ also satisfies the condition, which means there are infinitely many of them. □

11th Grade
First Day
(8:00 – 12:00; July 30, 2018)

1. In the coordinate plane, we call a point a rational point if both of its coordinates are rational numbers, and otherwise we call it an irrational point. Suppose there is a regular pentagon in the plane, then among its vertices, are there more rational points than irrational points or more irrational points than rational points? Prove your result. (Contributed by Wang Changping)

Solution We shall prove that for any regular pentagon in the coordinate plane, there are always more irrational points than rational points among its vertices.

In fact, let the side length and diagonal length of the pentagon be a, b, respectively. Then we have $\dfrac{a}{b} = \dfrac{\sqrt{5}-1}{2}$, so

$$\frac{a^2}{b^2} = \left(\frac{a}{b}\right)^2 = \left(\frac{\sqrt{5}-1}{2}\right)^2 = \frac{3-\sqrt{5}}{2},$$

which is irrational. Suppose there are three rational points in the vertices of the pentagon, then the pairwise distances of these vertices must contain both a and b. However, the square of the distance between any two rational points is a rational number, so both a^2 and b^2 are rational, which is contradictory to the fact that $\dfrac{a^2}{b^2}$ is irrational. Therefore, a regular pentagon in the coordinate plane cannot have more than 2 rational vertices, so it has more irrational vertices than rational vertices. $\qquad\square$

2 Let a be a real number, and sequence $a_1, a_2, \ldots$ satisfy:

$$a_1 = a, \quad a_{n+1} = \begin{cases} a_n - \dfrac{1}{a_n}, & a_n \neq 0 \\ 0, & a_n = 0 \end{cases} \quad (n = 1, 2, \ldots)$$

Find all real number a, such that $|a_n| < 1$ for any positive integer n. (Contributed by Li Shenghong)

Solution We discuss the following two cases.

Case 1: Some term in the sequence $\{a_n\}$ is 0. Let m be the smallest positive integer such that $a_m = 0$. If $m \geq 2$, then we have $a_{m-1} \neq 0$, which implies

$$a_{m-1} - \frac{1}{a_{m-1}} = a_m = 0.$$

Thus $a_{m-1} = \pm 1$, a contradiction. Hence the only possibility is $m = 1$, and in this case sequence $a_1 = a_2 = \cdots = 0$ satisfies the condition.

Case 2: For each positive integer n we have $a_n \in (-1, 0) \cup (0, 1)$. Since a_n and $\dfrac{1}{a_n}$ have the same sign, and $\dfrac{1}{|a_n|} > 1 > |a_n|$, we have

$$|a_{n+1}| = \left| a_n - \frac{1}{a_n} \right| = \frac{1}{|a_n|} - |a_n|.$$

Let $b_n = |a_n|$, then $b_n \in (0, 1)$ and $b_{n+1} = \dfrac{1}{b_n} - b_n$. Hence

$$b_{n+1} - \frac{\sqrt{2}}{2} = \frac{1}{b_n} - \sqrt{2} - b_n + \frac{\sqrt{2}}{2} = -\left(1 + \frac{\sqrt{2}}{b_n}\right)\left(b_n - \frac{\sqrt{2}}{2}\right).$$

Consequently, we have $\left| b_{n+1} - \dfrac{\sqrt{2}}{2} \right| = \left| b_1 - \dfrac{\sqrt{2}}{2} \right| \cdot \prod_{i=1}^{n} \left(1 + \dfrac{\sqrt{2}}{b_n} \right)$. If $b_1 \neq$

$\dfrac{\sqrt{2}}{2}$, then from

$$\prod_{i=1}^{n} \left(1 + \frac{\sqrt{2}}{b_n} \right) > (1 + \sqrt{2})^n \to +\infty \, (n \to \infty).$$

We see that $\{b_n\}$ is unbounded, which is a contradiction.

If $b_1 = \dfrac{\sqrt{2}}{2}$, then $a = \pm \dfrac{\sqrt{2}}{2}$, and the corresponding sequence is $a_n = (-1)^{n-1} a$, which satisfies the conditions.

In summary, all possible values of a are $0, \pm \dfrac{\sqrt{2}}{2}$. $\Box$

3   As shown in Fig. 3.1, $\triangle ABC$ is inscribed to $\odot O$, $\angle ABC > 90°$, M is the midpoint of BC, and P is a point inside the triangle such that $PB \perp PC$. Construct the line passing through P and perpendicular to AP, and let D, E be points on this line (which are different from P) such that $BD = BP, CE = CP$. Suppose that $ADOE$ is a parallelogram, prove that $\angle OPE = \angle AMB$. (Contributed by Lin Tianqi)

Solution   As shown in Fig. 3.2, let N be the intersection of AO and DE, which is the common midpoint of them. We first show that B, P, N, C are concyclic. Let B', C', M' be the projections of B, C, M on line DE. Then

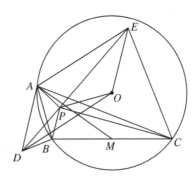

Fig. 3.1

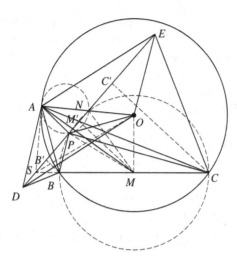

Fig. 3.2

they are the midpoints of $PD, PE, B'C'$, respectively. Thus

$$PB' = \frac{1}{2}DP = \frac{1}{2}DE - \frac{1}{2}PE = NE - C'E = NC'.$$

This means that M' is the midpoint of PN, i.e., $MP = MN$. From $PB \perp PC$ we see that $MB = MP = MC$, and P, N both lie on the circle with diameter BC.

Apparently P also lies on the circle with diameter AN, which is tangent to $\odot O$. We note that the tangent line to $\odot O$ at A, lines PN and BC are the three pairwise radical axes of circles $\odot O, \odot M, \odot AN$, so they are concurrent at the radical center S.

Since $\angle NAS = \angle APN = 90°$, we have

$$NO^2 = NA^2 = NP \cdot NS.$$

Hence $\triangle NOP \sim \triangle NSO$, and

$$\angle OPE = \angle OPN = \angle SON.$$

Also $\angle OAS = \angle OMS = 90°$, so S, A, O, M are concyclic, and

$$\angle SON = \angle SOA = \angle SMA = \angle AMB.$$

Therefore, $\angle OPE = \angle AMB.$ □

4 Determine whether exists set $A \subseteq \mathbb{N}^*$, such that for any positive integer n, $A \cap \{n, 2n, 3n, \ldots, 15n\}$ has exactly one element; and further, there are infinitely many positive integer m such that $\{m, m + 2018\} \subseteq A$. Prove your result. (Contributed by He Yijie)

Solution   We list the prime numbers greater than 2018 in the ascending order, as $p_1, p_2, \ldots$. Let N_0 denote the set of positive integers that do not have any prime factors greater than 2018, and for $i = 1, 2, \ldots$, also use N_i to denote the set of positive integers that do not have any prime factors greater than p_i. Apparently $N_0 \subseteq N_1 \subseteq N_2 \subseteq \cdots$ and $\bigcup_{i=0}^{\infty} N_i = \mathbb{N}^*$.

For any positive integer $l \in N_0$, if $l = 2^a 3^b 5^c 7^d 11^e 13^f t$ (where a, b, c, d, e, f are nonnegative integers and t is a positive integer that is not divisible by $2, 3, 5, 7, 11, 13$), then define

$$f_0(l) = a + 4b + 9c + 11d + 7e + 14f.$$

Inductively, for $i = 1, 2, \ldots$, we define functions f_i on sets N_i: let $l = 2^a 3^b 5^c 7^d 11^e 13^f t \cdot \prod_{j=1}^{i} p_j^{\lambda_j} \in N_i$, where $a, b, \ldots, f, \lambda_1, \ldots, \lambda_i \in N$ and t does not have any prime factor that is smaller than 15 or greater than 2018. Then

$$f_i(l) = a + 4b + 9c + 11d + 7e + 14f + \left(\sum_{j=1}^{i-1} \delta_j \lambda_j \right) + \delta_i \lambda_i, \qquad ①$$

where $\delta_i = f_{i-1}(p_i - 2018)$ (note that $p_i - 2018 \in N_{i-1}$, so $f_{i-1}(p_i - 2018)$ is already defined). From this definition, the restriction of f_i on N_{i-1} is exactly f_{i-1}, and

$$f_i(p_i) = \delta_i = f_{i-1}(p_i - 2018). \qquad ②$$

Thus, we may define function f on $\mathbb{N}^*$ as follows: choose i such that $l \in N_i$, and let $f(l) = f_i(l)$.

For any positive integer n, we prove that $f(kn)$ $(k = 1, 2, \ldots, 15)$ compose a complete residue system modulo 15. In fact, if $n = 2^a 3^b 5^c 7^d 11^e 13^f t \cdot \prod_{j=1}^{i} p_j^{\lambda_j} \in N_i$, where i is sufficiently large, then consider $\triangle_k = f(kn) - f(n)$ for $k = 1, 2, \ldots, 15$. Suppose the standard factorization of k is $2^{\alpha_1} 3^{\alpha_2} 5^{\alpha_3} 7^{\alpha_4} 11^{\alpha_5} 13^{\alpha_6}$, then

$$\triangle_k = f \left(2^{a+\alpha_1} 3^{b+\alpha_2} 5^{c+\alpha_3} 7^{d+\alpha_4} 11^{e+\alpha_5} 13^{f+\alpha_6} t \cdot \prod_{j=1}^{i} p_j^{\lambda_j} \right)$$

$$- f \left(2^a 3^b 5^c 7^d 11^e 13^f t \cdot \prod_{j=1}^{i} p_j^{\lambda_j} \right)$$

$$= \alpha_1 + 4\alpha_2 + 9\alpha_3 + 11\alpha_4 + 7\alpha_5 + 14\alpha_6.$$

Since

$$\triangle_1 = 0, \triangle_2 = 1, \triangle_3 = 4, \triangle_4 = 2, \triangle_5 = 9, \triangle_6 = 5, \triangle_7 = 11, \triangle_8 = 3,$$

$$\triangle_9 = 8, \triangle_{10} = 10, \triangle_{11} = 7, \triangle_{12} = 6, \triangle_{13} = 14, \triangle_{14} = 12, \triangle_{15} = 13,$$

it follows that $f(kn)$ $(k = 1, 2, \ldots, 15)$ compose a complete residue system modulo 15. From the definition of f, and combining with $\textcircled{2}$, we can also see that for any positive integer i, if $m_i = p_i - 2018$, then

$$f(m_i + 2018) = f_i(p_i) = f_{i-1}(p_i - 2018) = f(m_i).$$

By the pigeonhole principle, there are infinitely many $f(m_i)$ $(i \in \mathbb{N}^*)$ that are congruent modulo 15. Suppose the corresponding remainder is r, and let

$$A = \{n \in \mathbb{N}^* \,|\, f(n) \equiv r \,(\mathrm{mod}\,15)\,\}.$$

This set satisfies that $A \cap \{n, 2n, 3n, \ldots, 15n\}$ has exactly one element for every positive integer n, and there are infinitely many $i \in \mathbb{N}^*$ such that $\{m_i, m_i + 2018\} \subseteq A$.

In summary, a set with the desired properties exists. $\qquad\square$

Second Day
(8:00 – 12:00; July 31, 2018)

5 As shown in Fig. 5.1, in $\triangle ABC$, $AB = AC$. The center of $\odot O$ is the midpoint of BC, and it is tangent to AB, AC at E, F, respectively. Point G lies on $\odot O$, such that $AG \perp EG$. Tangent line to $\odot O$ at G intersects AC at K. Prove that line BK bisects segment EF. (Contributed by Zhang Pengcheng)

Solution As shown in Fig. 5.2, let D be the intersection of EF and BK. Extend KG to intersect AB at M. Since $ME = MG$ and $AG \perp EG$, M is the midpoint of AE. We apply Menelaus' theorem to triangle ABK and line EDF, and obtain

$$\frac{AE}{EB} \cdot \frac{BD}{DK} \cdot \frac{KF}{FA} = 1.$$

Since $AE = AF$, it follows that

$$\frac{BD}{BK} = \frac{BE}{FK}. \qquad\qquad \textcircled{1}$$

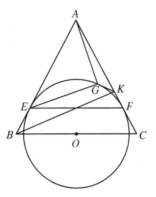

Fig. 5.1

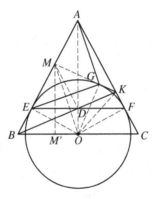

Fig. 5.2

Let M' be the projection of M on line BC. We have

$$\angle BOM = \angle BOE + \angle MOE = \angle BAO + \frac{1}{2}\angle EOG$$

$$= \frac{1}{2}\angle BAC + \frac{1}{2}\angle AMK = \frac{1}{2}\angle MKC = \angle CKO,$$

as well as $\angle B = \angle C$, so $\triangle BOM \sim \triangle CKO$, and MM' and OF are the corresponding altitudes. Hence,

$$\frac{BE}{FK} = \frac{CF}{FK} = \frac{BM'}{M'O} = \frac{BM}{MA}. \qquad \textcircled{2}$$

Combining ①, ② we obtain $\dfrac{BD}{DK} = \dfrac{BM}{MA}$, which means $MD \parallel AK$. Also M is the midpoint of AE, so MD is the median line of $\triangle AEF$, and D is the midpoint of EF. □

6 For integer $m \geq 2$, suppose there are $3m$ people attending a meeting, where each pair of them either shake hands once with each other, or do not shake hands. For a positive integer n ($n \leq 3m - 1$), if there is a collection of n people among them, whose number of hand-shakings are exactly $1, 2, \ldots, n$, then we say this meeting is "n-interesting". Suppose that for any possible meeting that is n-interesting, there always exist three attendees such that each pair of them have shaken hands with each other. Find the minimum of n. (Contributed by Zhang Pengcheng)

Solution Let the $3m$ people be $A_1, A_2, \ldots, A_{3m}$. If $1 \leq n \leq 2m$, then we construct the following n-interesting meeting: for $i = 1, 2, \ldots, m$, let A_{3m+1-i} shake hands with $A_i, A_{i+1}, \ldots, A_{2m}$, respectively.

Thus, the attendees $A_1, A_2, \ldots, A_m, A_{2m+1}, A_{2m+2}, \ldots, A_{3m}$ have shaken hands $1, 2, \ldots, 2m$ times, respectively. Since $2m \geq n$, this meeting is n-interesting. However, for any 3 attendees, at least two of them belong to either $S_1 = \{A_1, A_2, \ldots, A_{2m}\}$ or $S_2 = \{A_{2m+1}, A_{2m+2}, \ldots, A_{3m}\}$ simultaneously. From the construction, they have not shaken hands with each other. Hence, there do not exist 3 attendees having shaken hands with each other among them.

Thus, $n \geq 2m + 1$. Now for $n = 2m + 1$, we consider a meeting that is $(2m+1)$-interesting. There is someone who has shaken hands $2m+1$ times, who, we assume to be A_{3m}, has shaken hands with $A_1, A_2, \ldots, A_{2m+1}$. Let $T_1 = \{A_1, A_2, \ldots, A_{2m+1}\}$, $T_2 = \{A_{2m+2}, A_{2m+3}, \ldots, A_{3m}\}$.

Since the numbers of hand-shakings for people in T_2 take at most $m - 1$ different values, there exists $k \in \{m, m + 1, \ldots, 2m + 1\}$, such that no person in T_2 has shaken hands exactly k times. Hence someone in T_1 must have shaken hands k times, whom we assume to be A_i. Since A_i has shaken hands for fewer than $m - 1$ times with persons in T_2, there exists $A_j \in T_1$ ($A_j \neq A_i$) who has shaken hands with A_i. Thus, A_{3m}, A_i, A_j are the three attendees we look for.

In summary, any $(2m + 1)$-interesting meeting admits three attendees who have shaken hands with each other, so the minimum of n is $2m + 1$.

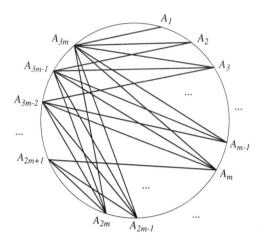

Fig. 6.1   ☐

7 For positive integers m, n, let $f(m, n)$ be the number of ordered integer tuples (x, y, z) that satisfy:
$$\begin{cases} xyz = x + y + z + m, \\ \max\{|x|, |y|, |z|\} \leq n. \end{cases}$$
Determine whether there exist m, n, such that $f(m, n) = 2018$, and prove your result. (Contributed by He Yijie)

Solution Suppose that $f(m, n) = 2018$ for positive integers m, n. Then for an integer solution (x, y, z) of
$$\begin{cases} xyz = x + y + z + m, \\ \max\{|x|, |y|, |z|\} \leq n, \end{cases} \qquad ①$$
we shall call it a type-1 solution if $x = y = z$, a type-2 solution if exactly two of them are equal, and a type-3 solution if no two of them are equal. For $i = 1, 2, 3$, let a_i be the number of type-i solutions. Since interchanging the values of x, y, z in any integer solution of ① still gives an integer solution, we have $3 \mid a_2, 6 \mid a_3$. Thus,
$$a_1 \equiv a_1 + a_2 + a_3 = 2018 \equiv 2 \pmod 3.$$
In other words, the number of integer roots of equation
$$x^3 - 3x - m = 0. \qquad ②$$
that satisfy $|x| < n$ is congruent to 2 modulo 3. In particular, there are at least 2 integer roots, which we denote as α, β. By Viete's theorem, the

third root γ satisfies

$$\begin{cases} \alpha + \beta + \gamma = 0, \\ \alpha\beta + \beta\gamma + \gamma\alpha = -3, \\ \alpha\beta\gamma = m, \end{cases}$$

Hence $\gamma = -\alpha - \beta$ is an integer. Without loss of generality assume $\alpha \leq \beta \leq \gamma$. Since $\alpha\beta\gamma = m > 0$, we have $\alpha \leq \beta < 0 < \gamma$. This implies

$$-3 = \alpha\beta + (\alpha + \beta)(-\alpha - \beta) = -\alpha^2 - \alpha\beta - \beta^2 \leq -3.$$

Therefore, $\alpha = \beta = -1$, and $\gamma = 2, m = 2$.

Now consider the equation for $m = 2$. Since there are only $(2n + 1)^3$ tuples (x, y, z) satisfying $\max\{|x|, |y|, |z|\} \leq n$, then from $(2n+1)^3 \geq 2018$, we get $n \geq 6$.

Now we calculate a_1, a_2, a_3.

(1) Since $-1, 2$ are the only integer solutions of $x^3 - 3x - 2 = 0, (-1, -1, -1)$ and $(2, 2, 2)$ are the only type-1 solutions of ①, so $a_1 = 2$.

(2) For type-2 solutions, suppose $y = z \neq x$. Then we have $xy^2 = x + 2y + 2$, so

$$(y + 1) \cdot (x(y - 1) - 2) = 0.$$

If $y = -1$, then the $2n$ tuples $(x, -1, -1)$ $(-n \leq x \leq n, x \neq -1)$ are all type-2 solutions of ①. If $y \neq -1$, then from $x(y - 1) = 2$ we have

$$\begin{cases} x = \pm 2, \pm 1, \\ y - 1 = \pm 1, \pm 2. \end{cases}$$

Since $x \neq y$, there are exactly two solutions $(x, y) = (-2, 0), (1, 3)$. Considering symmetry, there are $a_2 = 3(2n + 2)$ type-2 solutions.

(3) For type-3 solutions, we calculate the corresponding number of 3-element sets $\{x, y, z\}$. Note that $-1 \notin \{x, y, z\}$, since otherwise assuming $z = -1$, then $(x + 1)(y + 1) = 0$, which is contradictory to the assumption that no two of x, y, z are equal.

If $0 \in \{x, y, z\}$, assuming $z = 0$, then from ①, we have $x + y = -2$, and there are $n - 2$ such sets:

$$\{x, y, z\} = \{k, -2 - k, 0\} \, (k = 1, 2, \ldots, n - 2).$$

If $0 \notin \{x, y, z\}$, we assume y, z have the same sign, and $|y| \geq |z|$.

If $|z| = 1$, then necessarily $z = 1$, so that $(x - 1)(y - 1) = 4$. Since $x \neq y, xy \neq 0$, the only possibility is $(x, y) = (2, 5), (5, 2)$, in which case $\{x, y, z\} = \{1, 2, 5\}$.

If $|z| \geq 2$, then from $x = \dfrac{y + z + 2}{yz - 1}$ we obtain that

$$|x| \leq \frac{|y| + |z| + 2}{|y| \cdot |z| - 1} \leq \frac{2|y| + 2}{2|y| - 1} = 1 + \frac{3}{2|y| - 1} \leq 2.$$

Note that the equalities of the inequality signs above cannot hold simultaneously, because if they do, then $(x, y, z) = (2, 2, 2)$, which means it is not a type-3 solution. Hence $|x| < 2$, and $x = 1$, so we still have $\{x, y, z\} = \{1, 2, 5\}$.

Thus, the number of type-3 solutions is $a_3 = 6(n - 1)$.

Combining (1), (2), (3), we conclude that for every $n \geq 6$,

$$a_1 + a_2 + a_3 = 2 + 3(2n + 2) + 6(n - 1) = 12n + 2.$$

Since $a_1 + a_2 + a_3 = f(m, n) = 2018$, we get $n = 168$. Therefore, $f(m, n) = 2018$ if and only if $m = 2, n = 168$. □

Remark There is an alternative argument to prove that $m = 2$ after obtaining that ② has at least two distinct integer solutions. Consider function $\varphi(x) = x^3 - 3x$. Since

$$\varphi'(x) = 3x^2 - 3 \begin{cases} > 0, \ |x| > 1, \\ = 0, \ x = \pm 1, \\ < 0, \ |x| < 1, \end{cases}$$

it follows that $\varphi(x)$ is increasing on $(-\infty, -1]$ and $[1, +\infty)$, and decreasing on $[-1, 1]$. Note that $\varphi(-1) = 2$, so when $m \geq 3$, the equation $\varphi(x) - m = 0$ has exactly one real solution (which lies in the interval $[1, +\infty)$), which gives a contradiction.

If $m = 1$, then equation ② becomes $x^3 - 3x - 1 = 0$, which has no integer solutions since ± 1 are not its solutions. Therefore, we must have $m = 2$.

8 Suppose positive constant $C \geq 1$ and nonnegative real number sequence $\{a_n\}$ satisfies that for any real number $x \geq 1$,

$$\left| x \lg x - \sum_{k=1}^{[x]} \left[\frac{x}{k} \right] a_k \right| \leq Cx.$$

(Here $[\alpha]$ denotes the greatest integer that does not exceed α.) Prove that for every real number $y \geq 1$, we have

$$\sum_{k=1}^{[y]} a_k < 3Cy.$$

(Contributed by Li Shenghong)

Solution For any real number $x \geq \frac{1}{2}$, define

$$S(x) = \sum_{k=1}^{[x]} a_k, \quad T(x) = \sum_{k=1}^{[x]} \left[\frac{x}{k} \right] a_k.$$

For $\frac{1}{2} \leq x < 1$, we stipulate $S(x) = T(x) = 0$. Then we shall show that for $\lambda \geq \frac{1}{2}$,

$$S(2\lambda) - S(\lambda) \leq T(2\lambda) - 2T(\lambda). \qquad \text{(1)}$$

In fact, we note that $\left[\frac{2\lambda}{k} \right] = 1 \ (k = [\lambda] + 1, \ldots, [2\lambda])$, so

$$S(2\lambda) - S(\lambda) = \sum_{k=[\lambda]+1}^{[2\lambda]} a_k = \sum_{k=[\lambda]+1}^{[2\lambda]} \left[\frac{2\lambda}{k} \right] a_k$$

$$\leq \sum_{k=1}^{[\lambda]} \left(\left[\frac{2\lambda}{k} \right] - 2 \left[\frac{\lambda}{k} \right] \right) a_k + \sum_{k=[\lambda]+1}^{[2\lambda]} \left[\frac{2\lambda}{k} \right] a_k$$

$$= T(2\lambda) - 2T(\lambda).$$

Here $\sum_{k=1}^{[\lambda]} \left(\left[\frac{2\lambda}{k} \right] - 2 \left[\frac{\lambda}{k} \right] \right) a_k \geq 0$ because $\left[\frac{2\lambda}{k} \right] \geq 2 \left[\frac{\lambda}{k} \right], a_k \geq 0 (1 \leq k \leq [\lambda])$, and when $\frac{1}{2} \leq \lambda < 1$ it becomes an empty sum, and the same assertion still holds. Thus (1) holds.

Now consider any real number $y \geq 1$, and choose m to be the positive integer such that $2^{m-1} \leq y < 2^m$. From ① it follows that

$$S(y) = S(y) - S\left(\frac{y}{2^m}\right) = \sum_{k=1}^{m}\left(S\left(\frac{y}{2^{k-1}}\right) - S\left(\frac{y}{2^k}\right)\right)$$

$$\leq \sum_{k=1}^{m}\left(T\left(\frac{y}{2^{k-1}}\right) - 2T\left(\frac{y}{2^k}\right)\right) = T(y) - \left(\sum_{k=1}^{m-1}T\left(\frac{y}{2^k}\right)\right) - 2T\left(\frac{y}{2^m}\right)$$

$$= T(y) - \sum_{k=1}^{m-1}T\left(\frac{y}{2^k}\right).$$

Further, by the condition and the definition of $T(x)$, we see that $x\lg x - Cx \leq T(x) \leq x\lg x + Cx$. Hence,

$$S(y) \leq T(y) - \sum_{k=1}^{m-1}T\left(\frac{y}{2^k}\right) \leq y\lg y + Cy - \sum_{k=1}^{m-1}\left(\frac{y}{2^k}\lg\frac{y}{2^k} - \frac{Cy}{2^k}\right)$$

$$= y\lg y + Cy - \sum_{k=1}^{m-1}\frac{y}{2^k}\lg y + \sum_{k=1}^{m-1}\frac{y}{2^k}k\lg 2 + \sum_{k=1}^{m-1}\frac{Cy}{2^k}$$

$$= y\lg y\left(1 - \sum_{k=1}^{m-1}\frac{1}{2^k}\right) + Cy\left(1 + \sum_{k=1}^{m-1}\frac{1}{2^k}\right) + y\lg 2\sum_{k=1}^{m-1}\frac{k}{2^k}$$

$$< \frac{1}{2^{m-1}}y\lg y + 2Cy + 2\lg 2 \cdot y < 2\lg y + (2C + 2\lg 2)y.$$

Here the second last step uses the fact that $1 + \sum_{k=1}^{m-1}\frac{1}{2^k} < 2$ and

$$\sum_{k=1}^{m-1}\frac{k}{2^k} = 2\sum_{k=1}^{m-1}\frac{k}{2^k} - \sum_{k=1}^{m-1}\frac{k}{2^k}$$

$$= \sum_{k=0}^{m-2}\frac{k+1}{2^k} - \sum_{k=0}^{m-1}\frac{k}{2^k} = \left(\sum_{k=0}^{m-2}\frac{1}{2^k}\right) - \frac{m-1}{2^{m-1}} < 2.$$

The last step follows since $1 \leq 2^{m-1} \leq y < 2^m$.

Thus, it suffices to prove that $2\lg y + (2C + 2\lg 2)y < 3Cy$, or equivalently,

$$\frac{\lg y}{y} < \frac{C}{2} - \lg 2.$$

In fact, let $\lg y = t \geq 0$, then $y = 10^t = (1+9)^t \geq 1+9t$, and combining with $C \geq 1$, we have

$$\frac{\lg y}{y} < \frac{1}{9} < \frac{1}{2} - \lg 2 \leq \frac{C}{2} - \lg 2.$$

In summary, we conclude that $S(y) < 3Cy$ for all $y \geq 1$. $\square$

China Southeastern Mathematical Olympiad

2019 (Ji'an, Jiangxi)

From July 28 to August 2, 2019, the 16th China Southeastern Mathematical Olympiad (CSMO) was held at the Ji'an City No.1 Middle School in Jiangxi Province, China.

The main examination committee members of this competition are:

Tao Pingsheng (Jiangxi Science and Technology Normal University).

Li Shenghong (Zhejiang University);

Yang Xiaoming (Zhejiang University);

Liu Bin (Peking University);

Li Tiecheng (Tsinghua University);

Wu Quanshui (Fudan University);

Luo Ye (Hong Kong University);

Zhang Pengcheng (Fujian Normal University);

Dong Qiuixian (Nanchang University);

Wu Genxiu (Jiangxi Normal University);

Xiong Bin (East China Normal University);

He Yijie (East China Normal University).

The competition is divided into two levels: the first grade and the second grade of senior high school.

10th Grade
First Day
(8:00 – 12:00; July 30, 2019)

1 Find the maximal real number k, such that for all positive numbers a, b, the following inequality holds:

$$(a + b)(ab + 1)(b + 1) \geq kab^2.$$

(Contributed by Dong Qiuxian)

Solution 1 When $a = 1, b = 2$, we have $27 \geq 4k$, so $k \leq \dfrac{27}{4}$.

On the other hand, by the AM-GM inequality,

$$(a + b)(ab + 1)(b + 1) = \left(a + \frac{b}{2} + \frac{b}{2} \right) \left(\frac{ab}{2} + \frac{ab}{2} + 1 \right) \left(\frac{b}{2} + \frac{b}{2} + 1 \right)$$

$$\geq 3\sqrt[3]{a \cdot \frac{b}{2} \cdot \frac{b}{2}} \cdot 3\sqrt[3]{\frac{ab}{2} \cdot \frac{ab}{2} \cdot 1} \cdot 3\sqrt[3]{\frac{b}{2} \cdot \frac{b}{2} \cdot 1} = \frac{27}{4}ab^2.$$

Therefore, the maximum of k is $\dfrac{27}{4}$.

Solution 2 The problem is equivalent to finding the infimum of $\dfrac{(a + b)(ab + 1)(b + 1)}{ab^2}$ $(a, b > 0)$. First we have

$$\frac{(a + b)(ab + 1)(b + 1)}{ab^2} = \frac{b + 1}{b^2} \left(b^2 + 1 + b \left(a + \frac{1}{a} \right) \right)$$

$$\geq \frac{b + 1}{b^2} (b^2 + 1 + b \cdot 2) = \frac{(b + 1)^3}{b^2},$$

where the equality holds when $a = 1$. Hence, it is again equivalent to finding the infimum of $f(b) = \dfrac{(b + 1)^3}{b^2}$ $(b > 0)$. Since

$$f'(b) = \frac{(b + 1)(b - 2)}{b^3} \begin{cases} < 0, & 0 < b < 2, \\ = 0, & b = 2, \\ > 0, & b > 2, \end{cases}$$

it follows that $f(b)$ has minimum $f(2) = \dfrac{27}{4}$. Therefore, the maximum of k is $\dfrac{27}{4}$. $\qquad \square$

2 As shown in Fig. 2.1, two circles Γ_1, Γ_2 intersect at A, B. Let C, D be two points on Γ_1 and E, F be two points on Γ_2, such that A, B lie on segments CE, DF, respectively, and the two segments do not intersect. Segment CF intersect Γ_1, Γ_2 at $K(\neq C)$ and $L(\neq F)$, respectively, and segment DE intersects Γ_1, Γ_2 at $M(\neq D)$ and $N(\neq E)$, respectively. Prove that if the circumcircles of $\triangle ALM$ and $\triangle BKN$ are tangent to each other, then these two circles have the same radius. (Contributed by Zhang Pengcheng)

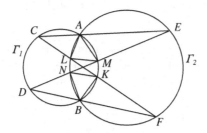

Fig. 2.1

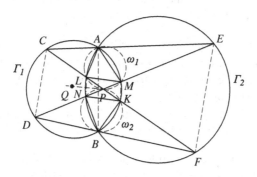

Fig. 2.2

Solution As shown in Fig. 2.2, the conditions imply that $\angle CDB = \angle BAE = 180° - \angle EFB$, so $CD//EF$. Thus, $\angle NLF = \angle NEF = \angle CDM = \angle CKM$, and $NL//MK$.

Let P be the intersection of CF, DE. Then

$$\angle ALP + \angle AMP = 180° - \angle AEF + 180° - \angle ACD = 180°.$$

So A, L, P, M are concyclic. Let ω_1 represent this circle. Then ω_1 is the circumcircle of $\triangle PLM$. Similarly, P, N, B, K are concyclic, and we denote this circle as ω_2, so that ω_2 is the circumcircle of $\triangle BKN$.

If ω_1, ω_2 are tangent to each other, then P is the point of tangency. Let PQ be the common tangent line to the two circles at P, where Q lies inside $\angle LPN$. Then $\angle LPN = \angle LPQ + \angle NPQ = \angle LMP + \angle NKP$. Since $\angle LPN = \angle LMP + \angle MLP$, we have $\angle NKP = \angle MLP$, which means $NK//LM$. Combining with the previous result that $NL//MK$, we see that $LMKN$ is a parallelogram, so ω_1, ω_2 have the same radius. $\qquad\Box$

3   Function $f : \mathbb{N}^* \to \mathbb{N}^*$ satisfies that for any positive integers a, b, $f(ab)$ divides $\max\{f(a), b\}$. Prove or disprove that there exist infinitely many positive integers k such that $f(k) = 1$. (Contributed by Luo Ye)

Solution   Let $p_1 < p_2$ be two prime numbers that are greater than $f(1)$. We prove by induction that $f(p_1^\alpha)\,|\,p_1$ for any integer $\alpha \geq 1$. We have $f(p_1)\,|\,\max\{f(1), p_1\} = p_1$, so $f(p_1) = 1$ or $f(p_1) = p_1$. Suppose that $f(p_1^\alpha)\,|\,p_1$ for $\alpha = k$. Then for $\alpha = k + 1$, since $f(p_1^{k+1})\,|\,\max\{f(p_1^k), p_1\}$ and $f(p_1^k)\,|\,p_1$, we get $f(p_1^{k+1})\,|\,p_1$. The induction is complete.

For any positive integer α such that $p_1^\alpha > p_2$, choose $x = p_1^\alpha p_2$. We have $f(x)\,|\,\max\{p_1^\alpha, f(p_2)\} = p_1^\alpha$ and $f(x)\,|\,\max\{p_2, f(p_1^\alpha)\} = p_2$. Thus, $f(x)\,|\,p_1$ and $f(x)\,|\,p_2$, so that $f(x) = 1$. Therefore, there are infinitely many x such that $f(x) = 1$. $\qquad\Box$

4   For a 2×5 rectangle on the grid table, placed either vertically or horizontally, we may delete one of the four unit squares on the corner, and call the shape formed by the remaining 9 squares a "flag", as shown in the following figure.

Given a 9×18 table, how many different ways are there if we want to cover the table with 18 "flags"? (Contributed by Tao Pingsheng)

Solution   For any $9 \times 2k$ grid table, where k is a positive integer, we call the side with length 9 the "base" of the table, and without loss of generality assume that the base is placed horizontally. We call a flag "horizontal",

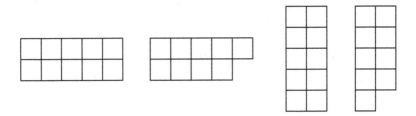

Fig. 4.1

if its side with length 5 is parallel to the base, and otherwise we call it "vertical".

In the following we shall say a region has type "$x-y-x$", if it is bounded from left, below and right by sides of lengths x, y, x, respectively, but not bounded above (as shown in the figure below). Also we call a region of type $x - y - x$ "uncoverable", if it is not possible to cover it with several flags (without overlapping).

First, it is easy to see that type 2-1-2 and type 2-3-2 are uncoverable, because, for type 2-1-2, no flag can reach the square at the bottom, and for type 2-3-2, the bottom side has to be covered by two vertical flags, which must overlap.

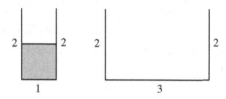

Next, type "4-5-4" is uncoverable: If the side with length 5 is covered by a horizontal flag, as shown in the figure below, then the deleted square has to be covered by a vertical flag, and the region on the left is then of type 2-3-2, which is uncoverable by the above argument. If the side with length 5 is covered by vertical flags, then at least 3 flags are required, while there are only 5 columns, so they must overlap. Hence, type 4-5-4 is uncoverable.

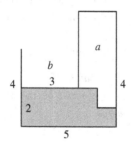

Further, type 4-7-4 is uncoverable: If the side with length 7 is contains the side of a horizontal flag, then the horizontal flag can be placed only in the following three ways (considering symmetry) and their flippings (for example, the flipping of A is D).

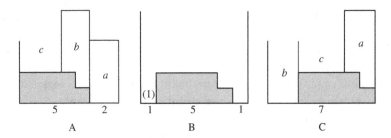

In Case A, regions a, b have to be covered by vertical flags, and region c is then of type 2-3-2.

In Case B, the left side has a region of type 2-1-2, which is uncoverable.

In Case C, regions a, b have to be covered by vertical flags, and region c is then of type 2-3-2.

If we flip the flags in Cases A, B, C, such as Case D in the following figure, then square (1) is uncoverable, so the whole region is uncoverable. Other cases are similar.

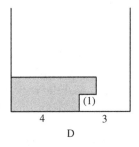

If the side with length 7 is not adjacent to any horizontal flag, then we need at least 4 vertical flags, which must overlap. Therefore, type 4-7-4 is also uncoverable.

Next we prove the following lemma.

Lemma *For every $k > 0$, if a table of size $9 \times 2k$ can be covered by $2k$ flags, then bottom 9×2 rectangle must be covered by two horizontal flags, in one of the following two ways*

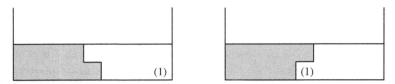

Proof of lemma We make induction on k. For $k = 1, 2$, all flags must be placed horizontally, so the base side is covered by two horizontal flags.

If $k \geq 3$, we consider the flag at the bottom left corner. There are several cases, as shown in the following.

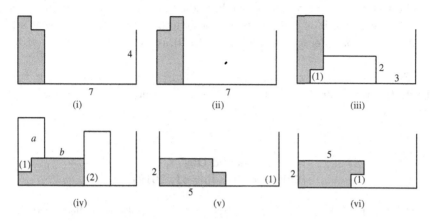

In Cases (i) and (ii), the regions in the right are of type 4-7-4, which is uncoverable.

In Case (iii), the white flag has to be placed as in the figure because of square (1). Then the right side has a region of type 2-3-2, which is uncoverable.

In Case (iv), flag a has to be placed in this way because of square (1). Then the square (2) cannot be covered by a vertical flag, since otherwise region b has type 2-3-2. Since (2) is also not coverable by a horizontal flag, this case is impossible.

In Case (v), the flag covering (1) have two possibilities. The left side is exactly the case we want, while on the right side, if we ignore the shaded flag, there is a region of type 4-7-4, which is uncoverable.

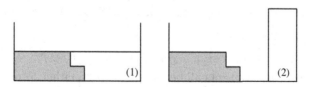

In Case (vi), the square (1) has to be covered by the white flag, which is exactly the case we want.

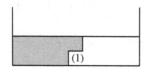

In summary, the bottom 9×2 rectangle is covered by 2 horizontal flags, and the lemma is proven.

From the lemma, we may prove by induction that in any covering of a $9 \times 2k$ rectangle, it is possible to divide it into k rectangles of size 9×2 (cut by lines parallel to the side of length 9), each of which is covered by two flags.

Since in every 9×2 rectangle there are 2 ways to place the two flags, the total number of different ways to cover a 9×18 rectangle is $2^9 = 512$. □

Second Day
(8:00 – 12:00; July 31, 2019)

5 A subset M of $S = \{1928, 1929, 1930, \ldots, 1949\}$ is called "red", if the sum of any two different elements in M is not divisible by 4. Let x be the number of 4-element red subsets of S, and y be the number of 5-element red subsets of S. Compare the sizes of x, y, and explain the reason. (Yang Xiaoming)

Solution We divide S into subsets S_0, S_1, S_2, S_3 based on the remainder on division by 4: If $x \in S$ and $x \equiv i \pmod 4$, then $x \in S_i$ ($i = 0, 1, 2, 3$). Apparently, $|S_0| = |S_1| = 6$, $|S_2| = |S_3| = 5$.

For any red subset M of S, we have $|M \cap S_0| \leq 1$, $|M \cap S_2| \leq 1$, and M cannot contain elements of S_1 and S_3 simultaneously (since otherwise there are two different elements in M whose sum is a multiple of 4). Conversely, if M satisfies the property above, then it is necessarily a red subset.

We first find the number x of different 4-element red subsets.

If $|M \cap S_0| = |M \cap S_2 0| = 0$, then $M \subseteq S_1$ or $M \subseteq S_3$, so there are $\binom{5}{4} + \binom{6}{4} = 20$ choices.

If $|M \cap S_0| = 1$, $|M \cap S_2| = 0$ or $|M \cap S_0| = 0$, $|M \cap S_2| = 1$, then the remaining three elements all belong to one part, either S_1 or S_3. Hence there are $(|S_0| + |S_2|) \times \left(\binom{6}{3} + \binom{5}{3} \right) = 330$ different choices of M.

If $|M \cap S_0| = |M \cap S_2| = 1$, then the other two elements belong to one of S_1 and S_3, so there are $|S_0| \times |S_2| \times \left(\binom{6}{2} + \binom{5}{2} \right) = 750$ different choices of M.

Thus, $x = 20 + 330 + 750 = 1100$.

With the same method we find the number of 5-element red subsets of S is

$$y = \left(\binom{6}{5} + \binom{5}{5}\right) + (|S_0| + |S_2|) \times \left(\binom{6}{4} + \binom{5}{4}\right)$$

$$+ |S_0| \times |S_2| \times \left(\binom{6}{4} + \binom{5}{3}\right)$$

$$= 7 + 220 + 900 = 1127.$$

Therefore, $x < y$. □

6 Let a, b, c be the side lengths of a given triangle. If x, y, z are positive real numbers such that $x + y + z = 1$, find the maximum of $axy + byz + czx$.

(Contributed by Li Shenghong)

Solution Without loss of generality, assume $0 < a \le b \le c$, and let

$$\begin{cases} p = \dfrac{a+c-b}{2}, \\ q = \dfrac{a+b-c}{2}, \\ r = \dfrac{b+c-a}{2}. \end{cases}$$

Then $p, q, r > 0$ and $\begin{cases} p + q = a, \\ q + r = b, \quad q \le p \le r. \text{ Thus,} \\ r + p = c, \end{cases}$

$$f(x, y, z) = (p + q)xy + (q + r)yz + (r + p)zx$$

$$= px(y + z) + qy(z + x) + rz(x + y)$$

$$= px(1 - x) + qy(1 - y) + rz(1 - z)$$

$$= \frac{p + q + r}{4} - \left[p\left(\frac{1}{2} - x\right)^2 + q\left(\frac{1}{2} - y\right)^2 + r\left(\frac{1}{2} - z\right)^2\right]$$

$$= \frac{p + q + r}{8} - \left[p\left(\frac{1}{2} - x\right)^2 + q\left(\frac{1}{2} - y\right)^2 + r\left(\frac{1}{2} - z\right)^2\right].$$

Let $m = \left| \dfrac{1}{2} - x \right|$, $n = \left| \dfrac{1}{2} - y \right|$, $l = \left| \dfrac{1}{2} - z \right|$, then $m, n, l \geq 0$, $m + n + l \geq \dfrac{1}{2}$. Now let $g(m, n, l) = pm^2 + qn^2 + rl^2$, and we proceed to find the minimum of $g(m, n, l)$. Since

$$\left(pm^2 + qn^2 + rl^2\right)\left(\frac{1}{p} + \frac{1}{q} + \frac{1}{r}\right) \geq (m + n + l)^2 \geq \frac{1}{4},$$

we have

$$pm^2 + qn^2 + rl^2 \geq \frac{pqr}{4(pq + qr + rp)},$$

where the equality holds when $pm = qn = rl$, or equivalently,

$$m = \frac{1}{2}\frac{qr}{pq + qr + rp}, \quad n = \frac{1}{2}\frac{pr}{pq + qr + rp}, \quad l = \frac{1}{2}\frac{pq}{pq + qr + rp}.$$

Therefore,

$$f(x, y, z) = \frac{a + b + c}{8} - g(m, n, l) \leq \frac{a + b + c}{8} - \frac{pqr}{4(pq + qr + rl)'},$$

where the equality is attained when $x = \dfrac{1}{2}\dfrac{pq + rp}{pq + qr + rp}$, $y = \dfrac{1}{2}\dfrac{pq + qr}{pq + qr + rp}$, $z = \dfrac{1}{2}\dfrac{qr + rp}{pq + qr + rp}$.

Now since $\begin{cases} p = \dfrac{a + c - b}{2}, \\ q = \dfrac{a + b - c}{2}, \\ r = \dfrac{b + c - a}{2}, \end{cases}$ it follows that

$$8pqr = (a + b - c)(a + c - b)(c + b - a).$$

Thus

$$4(pq + qr + rp) = (a + b - c)(a + c - b) + (c + b - a)(a + c - b)$$
$$+ (c + b - a)(a + b - c) = 2ab + 2bc + 2ca - a^2 - b^2 - c^2,$$

and

$$(a + b + c)(2ab + 2bc + 2ca - a^2 - b^2 - c^2) - 8pqr = 2abc,$$

$$f(x, y, z) \leq \frac{a + b + c}{8} - \frac{pqr}{4(pq + qr + rp)} = \frac{abc}{2ab + 2bc + 2ca - a^2 - b^2 - c^2}.$$

In summary, the maximum of $axy + byz + czx$ is $\dfrac{abc}{2ab + 2bc + 2ca - a^2 - b^2 - c^2}$. □

Remark We can also apply the Lagrange multiplier method. Construct Lagrange function $Q = axy + byz + czx + \lambda(x + y + z - 1)$. Calculating the partial derivatives, we have

$$
\begin{cases}
\dfrac{\partial Q}{\partial x} = ay + cz + \lambda = 0, \\[2mm]
\dfrac{\partial Q}{\partial y} = ax + bz + \lambda = 0, \\[2mm]
\dfrac{\partial Q}{\partial z} = by + cx + \lambda = 0, \\[2mm]
x + y + z = 1.
\end{cases}
$$

By solving the equations, we obtain that

$$
\begin{cases}
x = \dfrac{b(a + c - b)}{2ab + 2bc + 2ca - a^2 - b^2 - c^2}, \\[3mm]
y = \dfrac{c(a + b - c)}{2ab + 2bc + 2ca - a^2 - b^2 - c^2}, \\[3mm]
z = \dfrac{a(b + c - a)}{2ab + 2bc + 2ca - a^2 - b^2 - c^2}.
\end{cases}
$$

Since a, b, c are side lengths of a triangle, it follows that $x > 0, y > 0, z > 0$, so this is a valid solution.

In this case we have

$axy + byz + czx$

$$
= a\frac{b(a + c - b)}{2ab + 2bc + 2ca - a^2 - b^2 - c^2}\frac{c(a + b - c)}{2ab + 2bc + 2ca - a^2 - b^2 - c^2}
$$

$$
+ b\frac{c(a + b - c)}{2ab + 2bc + 2ca - a^2 - b^2 - c^2}\frac{a(b + c - a)}{2ab + 2bc + 2ca - a^2 - b^2 - c^2}
$$

$$
+ c\frac{a(b + c - a)}{2ab + 2bc + 2ca - a^2 - b^2 - c^2}\frac{b(a + c - b)}{2ab + 2bc + 2ca - a^2 - b^2 - c^2}
$$

$$
= \frac{abc}{2ab + 2bc + 2ca - a^2 - b^2 - c^2}.
$$

Therefore, the maximum of $axy + byz + czx$ is $\dfrac{abc}{2ab + 2bc + 2ca - a^2 - b^2 - c^2}$. $\qquad\square$

7 Let $ABCD$ be a given convex quadrilateral on the plane. Prove that there exist four different points P, Q, R, S on the same line, and square $A'B'C'D'$, such that P lies on lines $AB, A'B'$, Q lies on lines $BC, B'C'$, R lies on lines $CD, C'D'$ and S lies on lines $DA, D'A'$. (Contributed by He Yijie)

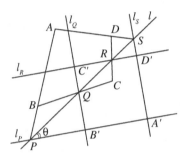

Fig. 7.1

Solution As shown in the figure, we construct line l that intersects the extension of AB at P, intersects sides BC, CD at Q, R, respectively, and intersects the extension of AD at S.

Suppose $QS = t \cdot PR$. Then we construct line l_P through P, such that the angle between l_P and l is θ, where $\tan\theta = t$. Let l_R be the line through R and parallel to l_P. Also construct lines l_Q, l_S through Q, S, respectively, which are perpendicular to l_P. Let A' be the intersection of l_P, l_S, B' be the intersection of l_P, l_Q, C' be the intersection of l_Q, l_R, and D' be the intersection of l_R, l_S. Then $A'B'C'D'$ is a rectangle. In addition, we have

$$\frac{A'B'}{C'D'} = \frac{QS \cdot \cos\theta}{PR \cdot \sin\theta} = \frac{t}{\tan\theta} = 1,$$

Hence, $A'B'C'D'$ is a square, and points P, Q, R, S and square $A'B'C'D'$ are the objects we desire. $\qquad\square$

Remark According to the solution above, the assumption that $ABCD$ is a convex is in fact unnecessary.

8 For a positive integer $x > 1$, define set

$S_x = \{p^\alpha \mid p \text{ is a prime factor of } x,\ 0 \le \alpha \le v_p(x),\ \alpha \equiv v_p(x) \ (\mathrm{mod}\,2)\}.$

Here $v_p(x)$ denotes the power of p in the standard factorization of positive integer x. Also let $f(x)$ be the sum of all the elements in S_x, and in addition define $f(1) = 1$.

For given positive integer m. Suppose positive integer sequence $a_1, a_2, \ldots, a_n, \ldots$ satisfies that for any integer $n > m$,

$$a_{n+1} = \max\{f(a_n), f(a_{n-1}+1), \ldots, f(a_{n-m}+m)\}.$$

(1) Prove that there exist constants $A, B\,(0 < A < 1)$, such that whenever x has at least two different prime factors, we have $f(x) < Ax + B$.

(2) Prove that there exists positive integer Q, such that for any $n \in N^*$, $a_n < Q$.

(Contributed by Luo Ye)

Solution Let $x = p_1^{\alpha_1} p_2^{\alpha_2} \cdots p_k^{\alpha_k}$, where $\alpha_1, \alpha_2, \ldots, \alpha_k$ are positive integers and $p_1 < p_2 < \cdots < p_k$ are prime numbers. Hence

$$f(x) = \sum_{i=1}^{k} \sum_{j=0}^{[\alpha_i/2]} p_i^{\alpha_i - 2j} \le \sum_{i=1}^{k} p_i^{\alpha_i} \frac{1}{1 - \frac{1}{p_i^2}}. \qquad \text{①}$$

We need the following lemma.

Lemma 1 *If x has at least two different prime factors, then $f(x) \le \frac{2}{3}(x+4)$.*

Proof of lemma 1 Let $x = p_1^{\alpha_1} p_2^{\alpha_2} \ldots p_k^{\alpha_k}$, then for $i = 1, 2, \ldots, k$,

$$f(x) = \sum_{i=1}^{k} \sum_{j=0}^{[\alpha_i/2]} p_i^{\alpha_i - 2j} \le \sum_{i=1}^{k} p_i^{\alpha_i} \frac{1}{1 - \frac{1}{p_i^2}} \le \sum_{i=1}^{k} \frac{4}{3} p_i^{\alpha_i}.$$

Thus, it suffices to show that $\frac{2}{3}\left(2\sum_{i=1}^{k} p_i^{\alpha_i} - \prod_{i=1}^{k} p_i^{\alpha_i}\right) \le \frac{8}{3}$.

Since $k \ge 2$, the left-hand side of the inequality above is decreasing in every variable $z_i = p_i^{\alpha_i}$. Thus,

$$\frac{2}{3}\left(2\sum_{i=1}^{k} p_i^{\alpha_i} - \prod_{i=1}^{k} p_i^{\alpha_i}\right) \le \frac{2}{3}(4k - 2^k).$$

If $k = 2$ or 3, then $\frac{2}{3}(4k - 2^k) = \frac{8}{3}$, while for $k \ge 4$, $\frac{2}{3}(4k - 2^k)$ is decreasing in k, so $\frac{2}{3}(4k - 2^k) \le \frac{2}{3}(4 \cdot 4 - 2^4) = 0$. Therefore, lemma 1 is proven.

Let $M_n = \max\{a_1, a_2, \ldots, a_n\}$, then $M(n)$ is nondecreasing in n. By lemma 1, for $x \geq 2m + 8$ and $j \in \{1, 2, \ldots, m\}$, if $x + j$ has at least two different prime factors, then $f(x + j) \leq \dfrac{2}{3}(x + m + 4) \leq x$.

If the sequence $a_1, a_2, \ldots$ is not bounded above, then there exists K such that for any $k > K$, we have $M(k) > 2m + 8$. Let $k_1 < k_2 < \cdots$ be the sequence of all integers k such that $k > K, M(k + 1) > M(k)$. Since $a_1, a_2, \ldots$ is not bounded above, sequence $\{k_n\}$ has infinitely many terms. For $j = 1, 2, \ldots$, since $M(k_j + 1) > M(k_j)$, there exists prime number p_j and positive integer α_j, such that $M(k_j + 1) = f(p_j^{\alpha_j})$. Let M_j denote $M(k_j + 1)$, and let $A_j = \dfrac{1}{1 - \dfrac{1}{p_j^2}}$. We know that for every $j > K + 1$,

$$M_j = f(p_j^{\alpha_j}) \leq p_j^{\alpha_j} \frac{1}{1 - \frac{1}{p_j^2}} \leq A_j(M_{j-1} + m). \qquad \textcircled{2}$$

Lemma 2   If $p_{l+1}, \ldots, p_{l+s}$ are different primes, then there exists a positive constant $C < 2$, such that $M_{t+s} \leq CM_l + Csm$.

Proof of lemma 2   Let $C = 3^{\frac{7}{12}} < 2$.

First we show that for any set of different primes $q_1, q_2, \ldots, q_r$ we have $\gamma = \prod_{i=1}^{r} \dfrac{1}{1 - \frac{1}{q_i^2}} < C$. This is because:

$$\ln \gamma = \sum_{i=1}^{r} \ln\left(1 + \frac{1}{q_i^2 - 1}\right) \leq \sum_{i=1}^{r} \frac{1}{q_i^2 - 1} \leq \frac{1}{2^2 - 1} + \sum_{\substack{q \text{ prime} \\ q \geq 3}} \frac{1}{q^2 - 1}$$

$$= \frac{1}{3} + \frac{1}{2} \cdot \sum_{\substack{q \text{ odd} \\ q \geq 3}} \left(\frac{1}{p - 1} - \frac{1}{p + 1}\right) = \frac{7}{12},$$

which means that $\gamma \leq e^{\frac{7}{12}} < 3^{\frac{7}{12}} < 2$.

Further, we know from (4) that $M_{j+s} \leq A_{j+s}(M_{j+s-1} + m)$ for $j = 1, 2, \ldots, l$. Applying this relation repeatedly, we obtain that

$$M_{l+s} \leq \prod_{j=1}^{m} A_{j+s} M_l + m \sum_{z=1}^{s} \prod_{j=1}^{z} A_{l+s+1-j}.$$

Since $p_{l+1}, \ldots, p_{l+s}$ are different, we have

$$\prod_{j=1}^{z} A_{l+s+1-j} \leq \prod_{j=1}^{m} A_{j+s} \leq C,$$

so $M_{l+s} \leq CM_l + Csm$, and lemma 2 is proven.

Now for any $j \geq K + 1$, if $p_{j+1}, p_{j+2}, \ldots$ are pairwise different prime numbers, the we can find a positive integer s such that $M_{s+j} \geq 2M_j$. Thus, $M_{j+s} \leq CM_j + Csm \leq \dfrac{C}{2}M_j + Csm$, so that $s \geq C_1 M_{j+s}$, where $C_1 = \dfrac{C}{\left(1 - \dfrac{C}{2}\right)m}$ is a positive constant. Since $p_{l+1}, \ldots, p_{l+s}$ are different prime numbers that are not greater than M_{s+j}, there are at least $C_1 M_{j+s}$ different prime numbers in $1, 2, \ldots, M_{j+s}$. However, the number of prime numbers not greater than M_{j+s} is close to $\dfrac{M_{j+s}}{\ln M_{j+s}} = o(M_{j+s})$ when M_{j+s} is large, so there is a contradiction.

If some prime number appears infinitely many times in $p_{K+1}, p_{K+2}, \ldots$, suppose $p_j = p_{j+s}$, and no two numbers in $p_{j+1}, \ldots, p_{j+s}$ are equal. Then there are infinitely many such pairs (j, s). Since $p_j = p_{j+s}$, we have

$$M_{j+s} = f(p_{j+s}^{\alpha_{j+s}}) > M_j = f(p_j^{\alpha_j}).$$

Hence, it follows that $\alpha_{j+s} \geq \alpha_j + 1$, and $M_{j+s} = f(p_{j+s}^{\alpha_{j+s}}) \geq p_j f(p_j^{\alpha_j}) \geq 2M_j$.

Thus, we still have $s \geq C_1 M_{j+s}$, which means that there are at least $C_1 M_{j+s}$ different prime numbers not greater than M_{j+s}, and this is impossible for large M_{j+s}.

Therefore, the sequence $a_1, a_2, \ldots$ is bounded above. Equivalently, there exists $Q > 0$ such that $a_k < Q$ for every positive integer k. $\qquad \square$

11th Grade
First Day
(8:00 – 12:00; July 30, 2019)

1 For any real number a, let $[a]$ denote the greatest integer that is not greater than a. Also let $\{a\} = a - [a]$. Determine whether there exist positive integers m, n and $n+1$ real numbers $x_0, x_1, \ldots, x_n$ such that

$$x_0 = 428, \quad x_n = 1928, \quad \frac{x_{k+1}}{10} = \left[\frac{x_k}{10}\right] + m + \left\{\frac{x_k}{5}\right\}$$

$$(k = 0, 1, \ldots, n - 1).$$

(Contributed by Wu Quanshui, He Yijie)

Solution Suppose such m, n and $x_0, x_1, \ldots, x_n$ exist. Let $y_k = \dfrac{x_k}{10}(k = 0, 1, \ldots, n)$. Then

$$y_0 = 42.8, y_n = 192.8, y_{k+1} = [y_k] + m + \{2y_k\}(k = 0, 1, \ldots, n - 1).$$

For $k = 0, 1, \ldots, n - 1$, since $[y_k] + m$ is an integer, and $0 \le \{2y_k\} < 1$, we have

$$[y_{k+1}] = [y_k] + m, \{y_{k+1}\} = \{2y_k\}.$$

Hence,

$$192 = [y_n] = [y_0] + mn = 42 + mn. \qquad \textcircled{1}$$

And $0.8 = \{y_n\} = \{2^n y_0\} = \{2^n \times 0.8\}$, or equivalently

$$\frac{2^n \times 8 - 8}{10} = \frac{4(2^n - 1)}{5} \in \mathbb{Z}. \qquad \textcircled{2}$$

Note that this is same as saying that $2^n \equiv 1 \pmod 5$, while this happens if and only if $4|n$. However, from $\textcircled{1}$ we see that $mn = 150$, so n is not a multiple of 4, which is a contradiction.

Therefore, there do not exist such m, n and $x_0, x_1, \ldots, x_n$. $\qquad \square$

2 As shown in Fig. 2.1, in parallelogram $ABCD$, $\angle BAD \neq 90°$. Suppose the circle centered at B with radius BA intersects the extensions of AB, CB at E, F, respectively. Also suppose the circle with center D and radius DA intersects the extensions of AD, CD at M, N, respectively. Lines EN, FM intersect at G, and lines AG, ME intersect at T. Line EN intersects $\odot D$ at $P(\neq N)$, and line MF intersects $\odot B$ at $Q(\neq F)$. Prove that G, P, T, Q are concyclic. (Contributed by Zhang Pengcheng)

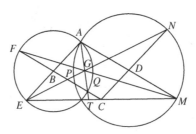

Fig. 2.1

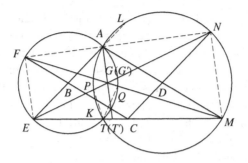

Fig. 2.2

Solution Suppose the second intersection of the circles centered at B and D is K. Then obviously E, K, M are collinear. Suppose the second intersection of EA and $\odot D$ is L. Then since $DN \parallel AL, AL \perp ML$, it follows that M is the midpoint of arc $\widehat{ML}$, so AN is the exterior angle bisector of $\angle EAM$. Similarly AF is also the exterior angle bisector of $\angle EAM$. So F, A, N are collinear.

Suppose the bisector of $\angle EAM$ intersects MF, ME at G', T', respectively. Then $AT' \perp FN$, and since $EF \perp FN, MN \perp FN$, we have $EF \parallel AT' \parallel MN$.

Hence,

$$\frac{AG'}{FE} = \frac{AN}{FN} \Rightarrow AG' = \frac{FE \cdot AN}{FN}, \qquad \text{(1)}$$

$$\frac{G'T'}{NM} = \frac{ET'}{EM} = \frac{AF}{FN} \Rightarrow G'T' = \frac{NM \cdot AF}{FN}. \qquad \text{(2)}$$

Combining with $\angle EAF = \angle MAN$, we derive that $\triangle AEF \sim \triangle AMN$, so that

$$\frac{AF}{AN} = \frac{FE}{NM} \Rightarrow FE \cdot AN = NM \cdot AF. \qquad \text{(3)}$$

By (1), (2), (3), we have $AG' = G'T'$, so MF passes through the midpoint of AT'. Similarly EN passes through the midpoint of AT' (which is G'). Therefore, G' coincides with G and T' coincides with T.

Since $AT \parallel MN$, we have $\angle GTK = \angle NMK = 180° - \angle GPK$, so G, P, K, T are concyclic. Further, since $AT \parallel EF$, we have $\angle QGT = \angle QFE = \angle QKT$, so G, K, T, Q are concyclic. Together, we conclude that G, P, T, Q are concyclic. □

3 In the first round, let n persons stand in a line, and label them with $1, 2, \ldots, n$ from left to right; then let those persons labeled with square numbers quit the line. In the second round, relabel the remaining people with $1, 2, \ldots$, preserving the original order, and let those persons labeled with square numbers quit.... Repeat this process, until all the persons quit the line. Let $f(n)$ denote the label of the person (in the first round) who quits the line in the last round, find a formula for $f(n)$. In particular, find the value of $f(2019)$. (Contributed by Tao Pingsheng)

Solution It is easy to find that $f(1) = 1$, $f(2) = 2$, $f(3) = f(4) = 3$, $f(5) = f(6) = 5$, $f(7) = f(8) = f(9) = 7$. According to these values, we guess that for integer m in the interval $[m^2 + 1, (m+1)^2]$,

$$f(n) = \begin{cases} m^2 + 1, & m^2 + 1 \le n \le m^2 + m \\ m^2 + m + 1, & m^2 + m + 1 \le n \le (m+1)^2 \end{cases} \qquad ①$$

In order to prove this assertion, we apply induction on m.

The case $1 \le n \le 3^2$ is already verified. Suppose the assertion holds for $n \in [m^2 + 1, (m+1)^2]$, consider the case for $n \in [(m+1)^2 + 1, (m+2)^2]$.

(1) If $(m+1)^2 + 1 \le n \le (m+1)^2 + (m+1)$, then after the first round, $m + 1$ people quit the line, and the person on the right end of the line is the one with label n in the first round.

The labels in the second round are $1, 2, \ldots, k$ where $k = n - (m+1)$, so that $m^2 + m + 1 \le k \le (m+1)^2$. By the induction hypothesis, the last one to quit the line is the one who has label $f(k)$ in the second round. For this person, the people who quit the line in the first round all stand on his/her left side, so this person should have $(m^2 + m + 1) + (m + 1) = (m+1)^2 + 1$ in the first round. Therefore, $f(n) = (m+1)^2 + 1$.

(2) If $(m+1)^2 + (m+1) + 1 \le n \le (m+2)^2$, then after the first round, the rightmost person A does not have label $(n+2)^2$ in the first round (the people with square labels have already quit the line). Thus, A has label $k + (m+1)$ in the first round (k is the number of people in the line after the first round), and $(m+1)^2 + (m+1) + 1 \le k + (m+1) \le (m+2)^2 - 1$, which means $(m+1)^2 + 1 \le k \le (m+1)^2 + (m+1)$. According to the induction hypothesis, the person who exits the line in the last round has label $f(k) = (m+1)^2 + 1$ in the second round. This person has label $(m+1)^2 + (m+1) + 1$ in the first round, so $f(n) = (m+1)^2 + (m+1) + 1$.

Thus, the induction is complete by (1), (2), and Formula ①️ is proven.

Note that $m < \sqrt{m^2 + 1} < \sqrt{m^2 + m} < m + \dfrac{1}{2} < \sqrt{m^2 + m + 1} < m+1$, so we eliminate m in ①️ and obtain

$$f(n) = \begin{cases} n - \sqrt{n} + 1, & n \in \mathbb{N}^* \\[2mm] [\sqrt{n}]^2 + 1, & 0 < \{\sqrt{n}\} < \dfrac{1}{2} \\[2mm] [\sqrt{n}]^2 + [\sqrt{n}] + 1, & \dfrac{1}{2} < \{\sqrt{n}\} < 1 \end{cases}$$

In particular, since $2019 \in [44 \times 45 + 1, 45^2]$, we have $f(2019) = 44 \times 45 + 1 = 1981$. □

Remark Formula ①️ can be compressed into a single expression

$$f(n) = \left[\frac{\sqrt{n-1} + n}{[\sqrt{n-1}] + 1} \right] \cdot [\sqrt{n-1}] + 1. \qquad ②️$$

We prove with following argument. By ①️ we have $m = [\sqrt{n-1}]$. Let

$$\alpha(n) = \begin{cases} 0, & m^2 + 1 \le n \le m^2 + m, \text{ (i.e., } 1 \le n - m^2 \le m), \\ 1, & m^2 + m + 1 \le n \le (m+1)^2, \text{ (i.e.,} \\ & m + 1 \le n - m^2 \le 2m + 1). \end{cases} \qquad ③️$$

Note that

$$\left[\frac{n - m^2}{m + 1} \right] = \begin{cases} 0, & m^2 + 1 \le n \le m^2 + m, \\ 1, & m^2 + m + 1 \le n \le (m+1)^2. \end{cases} \qquad ④️$$

So $\alpha(n) = \left[\dfrac{n - m^2}{m + 1} \right]$, and therefore,

$$f(n) = m^2 + 1 + \alpha(n) \cdot m = [\sqrt{n-1}]^2 + 1 + \left[\frac{n - [\sqrt{n-1}]^2}{[\sqrt{n-1}] + 1} \right] \cdot [\sqrt{n-1}]$$

$$= [\sqrt{n-1}]^2 + 1 + \left[\frac{n-1}{[\sqrt{n-1}] + 1} - [\sqrt{n-1}] + 1 \right] \cdot [\sqrt{n-1}]$$

$$= 1 + \left[\frac{n-1}{[\sqrt{n-1}] + 1} + 1 \right] \cdot [\sqrt{n-1}] = \left[\frac{\sqrt{n-1} + n}{[\sqrt{n-1}] + 1} \right] \cdot [\sqrt{n-1}] + 1$$

□

4   In a 5×5 matrix X, every entry is either 0 or 1. Let $x_{i,j}$ denote the entry in row i column j $(i, j = 1, 2, \ldots, 5)$. Consider the 5-number tuples in the columns, rows and diagonals of X:

$$(x_{i,1}, x_{i,2}, \ldots, x_{i,5}), (x_{i,5}, x_{i,4}, \ldots, x_{i,1}), \ (i = 1, 2, \ldots, 5),$$

$$(x_{1,j}, x_{2,j}, \ldots, x_{5,j}), (x_{5,j}, x_{4,j}, \ldots, x_{1,j}) \ (j = 1, 2, \ldots, 5),$$

$$(x_{1,1}, x_{2,2}, \ldots, x_{5,5}), (x_{5,5}, x_{4,4}, \ldots, x_{1,1}),$$

$$(x_{1,5}, x_{2,4}, \ldots, x_{5,1}), (x_{5,1}, x_{4,2}, \ldots, x_{1,5}).$$

If these ordered tuples are pairwise distinct, find all possible values of the sum of all the numbers in X. (Contributed by He Yijie)

Solution   Let Γ denote the set composed by all the 24 tuples in the problem. Then for each $\alpha = (a_1, a_2, \ldots, a_5) \in \Gamma$, we have $\beta = (a_5, a_4, \ldots, a_1) \in \Gamma$, and $\alpha \neq \beta$, so α is not palindromic. (Here a tuple (a, b, c, d, e) is palindromic if and only if $a = e$ and $b = d$.) The total number of 0-1 tuples that have length 5 and non-palindromic is $2^5 - 2^3 = 24$, so they are exactly all the elements in Γ.

For $\alpha \in \Gamma$, let S_α denote the sum of all the terms in α. Consider $\sum_{\alpha \in \Gamma} S_\alpha$.

On one hand, each $\alpha = (a_1, a_2, \ldots, a_5) \in \Gamma$ is paired up with $\alpha' = (1 - a_1, 1 - a_2, \ldots, 1 - a_5) \in \Gamma$. Note that $a_3 \neq 1 - a_3$, so $\alpha \neq \alpha'$. Hence, the elements in Γ can be divided into 12 pairs, where in each pair we have $S_\alpha + S_{\alpha'} = 5$. This implies $\sum_{\alpha \in \Gamma} S_\alpha = 60$.

On the other hand, in the sum $\sum_{\alpha \in \Gamma} S_\alpha$, the central entry $x_{3,3}$ in X is counted 8 times, the other diagonal entries $x_{1,1}, x_{2,2}, x_{4,4}, x_{5,5}, x_{1,5}, x_{2,4}, x_{4,2}, x_{5,1}$ are counted 6 times, and non-diagonal entries are counted 4 times. Therefore, we have

$$8x_{3,3} + 6(x_{1,1} + x_{2,2} + x_{4,4} + x_{5,5} + x_{1,5} + x_{2,4} + x_{4,2} + x_{5,1})$$

$$+ 4(x_{1,2} + x_{1,3} + \cdots + x_{5,4}) = 60.$$

Let S be the sum of all the numbers in X, then the equation above becomes

$$2x_{3,3} + (x_{1,1} + x_{2,2} + x_{4,4} + x_{5,5} + x_{1,5} + x_{2,4} + x_{4,2} + x_{5,1}) + 2S = 30. \quad \text{(1)}$$

Note that Equation ① immediately implies that $S \leq 15$.

(1) If $S = 15$, then by ① we see that all the diagonal entries are 0, in which case the diagonal tuple is palindromic, and this is impossible.

(2) If $S = 14$, then consider $x_{3,3}$. If $x_{3,3} = 1$, then the other diagonal entries are all 0, and the diagonal tuple is stil l palindromic, which is impossible.

If $x_{3,3} = 0$, then exactly two numbers in $x_{1,1}, x_{2,2}, x_{4,4}, x_{5,5}$, $x_{1,5}, x_{2,4}, x_{4,2}, x_{5,1}$ are 1, and they cannot belong to the same diagonal, since otherwise the other diagonal is palindromic. In addition, the case $x_{1,1} = x_{1,5} = x_{5,1} = x_{5,5} = 0$ is impossible, since $(0,0,0,1,0)$ will appear twice, and similarly it is impossible that $x_{2,2} = x_{2,4} = x_{4,2} = x_{4,4} = 0$. Thus, by symmetry we assume that

$$X = \begin{pmatrix} 1 & * & * & * & 0 \\ * & 0 & * & 1 & * \\ * & * & 0 & * & * \\ * & 0 & * & 0 & * \\ 0 & * & * & * & 0 \end{pmatrix}$$

Consider the tuples in Γ where 1 appears 3 or 4 times (there are 12 such tuples). They can only appear in columns 1, 2, 3 or rows 1, 3, 4 of X (note that they are not palindromic). Hence, the tuples in columns 1, 2, 3 and rows 1, 3, 4 (and there are $6 \times 2 = 12$ such tuples) are exactly these tuples. However, none of columns 4, 5, rows 2, 5 and the diagonals have tuple $(0,0,0,1,1)$, so $(0,0,0,1,1) \notin \Gamma$, which is a contradiction.

(3) If $S = 13$, then such matrix X exists, as shown below:

$$X = X_0 = \begin{pmatrix} 1 & 0 & 0 & 0 & 0 \\ 1 & 1 & 1 & 1 & 0 \\ 1 & 1 & 0 & 0 & 1 \\ 0 & 0 & 1 & 0 & 1 \\ 1 & 1 & 1 & 0 & 0 \end{pmatrix}$$

We can easily verify that this matrix satisfies thet condition.

(4) If $S \leq 12$, then change every term $x_{i,j}$ in X into $1 - x_{i,j}$, and let X' denote the new matrix. It is easy to see that X' also satisfies the condition, and the sum of numbers in X' is $S' = 25 - S \geq 13$. From the analysis above, we see that $S' = 13$, so $S = 12$. The example for this case is simply exchanging the 0 and 1 in X_0.

In summary, the sum of all the numbers in X can be 13 or 12. $\square$

Second Day
(8:00 – 12:00; July 31, 2019)

5 For any positive integer n, let a_n denote the number of different triangles with integer side lengths, whose longest side has length $2n$.

(1) Find the general formula for a_n;

(2) Suppose sequence $\{b_n\}$ satisfies that $\sum_{k=1}^{n}(-1)^{n-k}C_n^k b_k = a_n (n \in \mathbb{N}^*)$. Find the number of different positive integers n such that $b_n \leq 2019a_n$. (Contributed by Wu Genxiu)

Solution (1) Let $a_n(k)$ denote the number of triangles with integer side lengths, whose longest side has length $2n$ and whose shortest side has length k, then $a_n = \sum_{k=1}^{2n} a_n(k)$.

If $k \leq n$, then the length of the third side can be $2n+1-k, 2n+2-k, \ldots, 2n$, where there are k different choices. If $n < k \leq 2n$, then the length of the third side can be $k, k+1, \ldots, 2n$, where there are $2n+1-k$ different choices. Hence,

$$a_n(k) = \begin{cases} k, & 1 \leq k \leq n, \\ 2n+1-k, & n < k \leq 2n. \end{cases}$$

Consequently,

$$a_n = \sum_{k=1}^{2n} a_n(k) = \sum_{k=1}^{n} k + \sum_{k=n+1}^{2n}(2n+1-k) = n(n+1).$$

(2) Since $\sum_{k=1}^{n}(-1)^{n-k}C_n^k b_k = a_n$, we have

$$\sum_{k=1}^{n} C_n^k a_k = \sum_{k=1}^{n} C_n^k \left(\sum_{t=1}^{k}(-1)^{k-t}C_k^t b_t \right)$$

$$= \sum_{t=1}^{n} \left(\sum_{k=1}^{n}(-1)^{k-t}C_n^k C_k^t \right) b_t = b_n.$$

The last step follows because $\sum_{k=1}^{n}(-1)^{k-t}C_n^k C_k^t = \begin{cases} 1, & n = t \\ 0, & n > t \end{cases}.$

Equivalently,

$$b_n = \sum_{k=1}^{n} C_n^k a_k = \sum_{k=1}^{n} C_n^k k(k+1)$$

$$= n(n+1) \cdot 2^{n-2} + n \cdot 2^{n-1} = n(n+3) \cdot 2^{n-2}.$$

If $\dfrac{b_n}{a_n} \leq 2019$, then $n(n+3)2^{n-2} \leq 2019n(n+1)$. When $n \geq 2$, we have

$$\frac{n(n+3)}{n(n+1)}2^{n-2} = \frac{n+3}{n+1}2^{n-2} \leq 2019 \Rightarrow 2^{n-2} \leq 2019 \Rightarrow n \leq 12,$$

and $n = 1, 2, \ldots, 12$ all satisfy the inequality above. Therefore, the number of n such that $\dfrac{b_n}{a_n} \leq 2019$ is 12. $\qquad \square$

6 In $\triangle ABC$, $AB > AC$, the angle bisector of $\angle ABC$ intersects AC at D, and the angle bisector of $\angle ACB$ intersects AB at E. The line through A tangent to the circumcircle of $\triangle ABC$ intersects line ED at P, and suppose $AP = BC$. Prove that $BD \parallel CP$. (Contributed by Zhang Pengcheng)

Solution Let I be the incenter of $\triangle ABC$, (or equivalently, I is the intersection of BD, CE) and suppose lines BD, CE intersect the circumcircle of $\triangle ABC$ again at Q, R, respectively. From the property of the incenter we have $RI = RA, QI = QA$, so RQ is the perpendicular bisector of AI.

We shall prove that P, Q, R are collinear. This is equivalent to

$$\frac{EP}{PD} \cdot \frac{DQ}{QI} \cdot \frac{IR}{RE} = 1.$$

Let $\alpha = \dfrac{1}{2}\angle BAC$, $\beta = \dfrac{1}{2}\angle ABC$, $\gamma = \dfrac{1}{2}\angle ACB$, then $\alpha + \beta + \gamma = 90°$. It is simple to see that $\angle DAQ = \beta$, $\angle IAQ = \alpha + \beta$, $\angle EAR = \gamma$, $\angle IAR = \alpha + \gamma$,

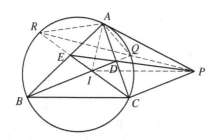

Fig. 6.1

and since AP is tangent to the circumcircle, we also have $\angle DAP = 2\beta$, $\angle EAP = 180° - 2\gamma$. Thus,

$$\frac{EP}{PD} \cdot \frac{DQ}{QI} \cdot \frac{IR}{RE} = \frac{S_{\triangle EAP}}{S_{\triangle DAP}} \cdot \frac{S_{\triangle DAQ}}{S_{\triangle IAQ}} \cdot \frac{S_{\triangle IAR}}{S_{\triangle EAR}}$$

$$= \frac{AE \cdot \sin\angle EAP}{AD \cdot \sin\angle DAP} \cdot \frac{AD \cdot \sin\angle DAQ}{AI \cdot \sin\angle IAQ} \cdot \frac{AI \cdot \sin\angle IAR}{AE \cdot \sin\angle EAR}$$

$$= \frac{\sin(180° - 2\gamma)}{\sin 2\beta} \cdot \frac{\sin\beta}{\sin(\alpha+\beta)} \cdot \frac{\sin(\alpha+\gamma)}{\sin\gamma}$$

$$= \frac{\sin 2\gamma}{\sin 2\beta} \cdot \frac{\sin\beta}{\cos\gamma} \cdot \frac{\cos\beta}{\sin\gamma} = 1.$$

Therefore, P, Q, R are collinear. Note that since RQ is the perpendicular bisector of AI, by symmetry we have

$$IP = AP = BC, \quad \angle QIP = \angle QAP = \angle QBC,$$

so $IP \parallel BC$ and $IP = BC$, which means $BCPI$ is a parallelogram, and $BD \parallel CP$.   □

⑦ Ann and Bob play a game, in which they take turns to choose numbers in $0, 1, \ldots, 81$, without repetition (each turn one chooses a number that has not been chosen in previous turns), and Ann starts first. When the 82 numbers are all chosen, let A denote the sum of all the numbers Ann chooses, and B denote the sum of all the numbers Bob chooses. In this game, Ann's goal is to make $\gcd(A, B)$ as large as possible, while Bob's goal is to make $\gcd(A, B)$ as small as possible. Find the value of $\gcd(A, B)$ after the game, under the assumption that both players adopt the best possible strategy. (Contributed by Luo Ye)

Solution   Let $d = \gcd(A, B)$. Note that $d | (A + B)$, where $A + B = 0 + 1 + \cdots + 81 = 81 \times 41$, so we have $d | 81 \times 41$.

We first show that Ann can ensure that $41 | d$. Divide the numbers $0, 1, \ldots, 81$ into 41 pairs:

$$\{x, x + 41\}(x = 0, 1, \ldots, 40).$$

Ann can make sure that she chooses exactly one number in every pair. In fact, except for the last turn, suppose Bob has chosen number a in the previous turn, then Ann considers number a' that matches a in the pairing above. If a' is not already chosen, then Ann chooses a', while if a' is already chosen, then Ann chooses an arbitrary number in the remaining numbers. Thus, before Bob chooses the last number, Ann has chosen one number out of every pair, so her purpose is reached.

In this case, $A \equiv 0 + 1 + \cdots + 40 \equiv 0 \pmod{41}$, and $B \equiv 0 \pmod{41}$ as well, so $41 | d$.

Next, we show that Bob can make sure that d is not a multiple of 3, so that $d | 41$.

We also divide $0, 1, \ldots, 81$ into 41 pairs:

$$\{x, x + 39\} \ (x = 0, 1, \ldots, 38), \{78, 81\}\{79, 80\}.$$

Then Bob can choose the number that matches the number Ann has chosen in the previous turn, so he can choose exactly one number out of every pair. However, in this case the two numbers in each pair differ by a multiple of 3, except for the last pair $\{79, 80\}$. Therefore, $A - B$ is not a multiple of 3, so d is not a multiple of 3. Combining with $d | 81 \times 41$, we conclude that $d = 41$.

In summary, the outcome of $\gcd(A, B)$ will be 41 if both players apply the best strategy. □

8 For positive integer $x > 1$, define

$$S_x = \{p^\alpha \,|\, p \text{ is a prime factor of } x, 0 \le \alpha \le v_p(x), \alpha \equiv v_p(x) \pmod 2\}.$$

Here $v_p(x)$ denotes the power of p in the standard factorization of x. Also let $f(x)$ be the sum of all the elements in S_x, and in addition define $f(1) = 1$.

Now given positive integer m, suppose positive integer sequence $a_1, a_2, \ldots, a_n, \ldots$ satisfies that for any integer $n > m$, we have $a_{n+1} = \max\{f(a_n), f(a_{n-1} + 1), \ldots, f(a_{n-m} + m)\}$.

(1) Prove that there exist constants $A, B(0 < A < 1)$, such that whenever x has at least two different prime factors, we have $f(x) < Ax + B$.

(2) Prove that there exist positive integers N, l, such that for any $n \ge N$, we have $a_{n+l} = a_n$.

(Contributed by Luo Ye)

Solution   Let $x = p_1^{\alpha_1} p_2^{\alpha_2} \cdots p_k^{\alpha_k}$, where $\alpha_1, \alpha_2, \ldots, \alpha_k$ are positive integers and $p_1 < p_2 < \cdots < p_k$ are prime numbers. Hence

$$f(x) = \sum_{i=1}^{k} \sum_{j=0}^{[\alpha_i/2]} p_i^{\alpha_i - 2j} \le \sum_{i=1}^{k} p_i^{\alpha_i} \frac{1}{1 - \frac{1}{p_i^2}}. \qquad ①$$

We need the following lemma.

Lemma 1   *If x has at least two different prime factors, then $f(x) \le \dfrac{2}{3}(x+4)$.*

Proof of lemma 1   Let $x = p_1^{\alpha_1} p_2^{\alpha_2} \cdots p_k^{\alpha_k}$, then for $i = 1, 2, \ldots, k$,

$$f(x) = \sum_{i=1}^{k} \sum_{j=0}^{[\alpha_i/2]} p_i^{\alpha_i - 2j} \le \sum_{i=1}^{k} p_i^{\alpha_i} \frac{1}{1 - \frac{1}{p_i^2}} \le \sum_{i=1}^{k} \frac{4}{3} p_i^{\alpha_i}.$$

Thus, it suffices to show that $\dfrac{2}{3}(2 \sum_{i=1}^{k} p_i^{\alpha_i} - \prod_{i=1}^{k} p_i^{\alpha_i}) \le \dfrac{8}{3}$. Since $k \ge 2$, the left-hand side of the inequality above is decreasing in every variable $z_i = p_i^{\alpha_i}$. Thus,

$$\frac{2}{3}\left(2 \sum_{i=1}^{k} p_i^{\alpha_i} - \prod_{i=1}^{k} p_i^{\alpha_i}\right) \le \frac{2}{3}(4k - 2^k).$$

If $k = 2$ or 3, then $\dfrac{2}{3}(4k - 2^k) = \dfrac{8}{3}$, while for $k \ge 4$, $\dfrac{2}{3}(4k - 2^k)$ is decreasing in k, so $\dfrac{2}{3}(4k - 2^k) \le \dfrac{2}{3}(4 \cdot 4 - 2^4) = 0$. Therefore, lemma 1 is proven.

Let $M_n = \max\{a_1, a_2, \ldots, a_n\}$, then $M(n)$ is nondecreasing in n. By lemma 1, for $x \ge 2m + 8$ and $j \in \{1, 2, \ldots, m\}$, if $x + j$ has at least two different prime factors, then $f(x+j) \le \dfrac{2}{3}(x + m + 4) \le x$.

If sequence $a_1, a_2, \ldots$ is not bounded above, then there exists K such that for any $k > K$, we have $M(k) > 2m + 8$. Let $k_1 < k_2 < \cdots$ be the sequence of all the integers k such that $k > K, M(k + 1) > M(k)$. Since $a_1, a_2, \ldots$ is not bounded above, sequence $\{k_n\}$ has infinitely many terms. For $j = 1, 2, \ldots$, since $M(k_j + 1) > M(k_j)$, there exists prime number p_j and positive integer α_j, such that $M(k_j + 1) = f(p_j^{\alpha_j})$. Let M_j denote $M(k_j + 1)$, and let $A_j = \dfrac{1}{1 - \dfrac{1}{p_j^2}}$. We know that for each $j > K + 1$,

$$M_j = f(p_j^{\alpha_j}) \le p_j^{\alpha_j} \frac{1}{1 - \frac{1}{p_j^2}} \le A_j(M_{j-1} + m). \qquad \text{(2)}$$

Lemma 2   *If $p_{l+1}, \ldots, p_{l+s}$ are different primes, then there exists positive constant $C < 2$, such that $M_{t+s} \le C M_l + C s m$.*

Proof of lemma 2   Let $C = 3^{\frac{7}{12}} < 2$.

First we show that for any set of different primes $q_1, q_2, \ldots, q_r$ we have $\gamma = \prod_{i=1}^{r} \dfrac{1}{1 - \dfrac{1}{q_i^2}} < C$. This is because:

$$\ln \gamma = \sum_{i=1}^{r} \ln\left(1 + \frac{1}{q_i^2 - 1}\right) \le \sum_{i=1}^{r} \frac{1}{q_i^2 - 1} \le \frac{1}{2^2 - 1} + \sum_{\substack{q \text{ prime} \\ q \ge 3}} \frac{1}{q^2 - 1}$$

$$= \frac{1}{3} + \frac{1}{2} \cdot \sum_{\substack{q \text{ odd} \\ q \ge 3}} \left(\frac{1}{p - 1} - \frac{1}{p + 1}\right) = \frac{7}{12},$$

which means that $\gamma \le e^{\frac{7}{12}} < 3^{\frac{7}{12}} < 2$.

Further, we know from (4) that $M_{j+s} \le A_{j+s}(M_{j+s-1} + m)$ for $j = 1, 2, \ldots, l$. Applying this relation repeatedly, we obtain that

$$M_{l+s} \le \prod_{j=1}^{m} A_{j+s} M_l + m \sum_{z=1}^{s} \prod_{j=1}^{z} A_{l+s+1-j}.$$

Since $p_{l+1}, \ldots, p_{l+s}$ are different, we have

$$\prod_{j=1}^{z} A_{l+s+1-j} \le \prod_{j=1}^{m} A_{j+s} \le C,$$

so $M_{l+s} \le CM_l + Csm$, and lemma 2 is proven.

Now for any $j \ge K + 1$, if $p_{j+1}, p_{j+2}, \ldots$ are pairwise different prime numbers, then we can find a positive integer s such that $M_{s+j} \ge 2M_j$. Thus, $M_{j+s} \le CM_j + Csm \le \dfrac{C}{2} M_j + Csm$, so that $s \ge C_1 M_{j+s}$, where $C_1 = \dfrac{C}{\left(1 - \dfrac{C}{2}\right) m}$ is a positive constant. Since $p_{l+1}, \ldots, p_{l+s}$ are different prime numbers that are not greater than M_{s+j}, there are at least $C_1 M_{j+s}$ different prime numbers in $1, 2, \ldots, M_{j+s}$. However, the number of prime numbers not greater than M_{j+s} is close to $\dfrac{M_{j+s}}{\ln M_{j+s}} = o(M_{j+s})$ when M_{j+s} is large, so there is a contradiction.

If some prime number appears infinitely many times in $p_{K+1}, p_{K+2}, \ldots$, suppose $p_j = p_{j+s}$, and no two numbers in $p_{j+1}, \ldots, p_{j+s}$ are equal. Then there are infinitely many such pairs (j, s). Since $p_j = p_{j+s}$, we have

$$M_{j+s} = f(p_{j+s}^{\alpha_{j+s}}) > M_j = f(p_j^{\alpha_j}).$$

Hence, it follows that $\alpha_{j+s} \ge \alpha_j + 1$, and $M_{j+s} = f(p_{j+s}^{\alpha_{j+s}}) \ge p_j f(p_j^{\alpha_j}) \ge 2M_j$.

Thus, we still have $s \geq C_1 M_{j+s}$, which means that there are at least $C_1 M_{j+s}$ different prime numbers not greater than M_{j+s}, and this is impossible for large M_{j+s}.

Hence, sequence $a_1, a_2, \ldots$ is bounded above, and there exists $Q > 0$, such that $a_k < Q$ for eachy positive integer k. Note that this implies that tuple $(a_k, a_{k+1}, \ldots, a_{k+m})$ has only finitely many choices, so there exist positive integers N, l such that $(a_N, a_{N+1}, \ldots, a_{N+m}) = (a_{N+l}, a_{N+l+1}, \ldots, a_{N+m+l})$. Then from the recurrence relation in the problem we have $(a_n, a_{n+1}, \ldots, a_{n+m}) = (a_{n+l}, a_{n+l+1}, \ldots, a_{n+m+l})$ for any $n \geq N$, and the desired proposition is proven. $\square$

International Mathematical Olympiad

The 60th International Mathematical Olympiad (IMO) was held in Bath, United Kingdom from July 11 to 22, 2019, Brazilian time. 621 students from 112 countries and regions participated in the competition. The Chinese team won the first place in the team total score with 227 points (tied with the US team), and all 6 players won gold medals.

The members of the Chinese team are as follows:

Team Leader: Xiong Bin, East China Normal University

Deputy Leader: He Yijie, East China Normal University

Observers: Qu Zhenhua, East China Normal University;
Wang Guangting, Shanghai Middle School in Shanghai;
Zhang Zeng, Youth Science and Technology Center of China Association for Science and Technology

Team Contestants:

Deng Mingyang, First Grade Student from Middle School Affiliated to Renmin University of China; 35 points, Gold Medal

Hu Sulin, Second Grade Student from Middle School Affiliated to South China Normal University; 39 points, Gold Medal

Xie Baiting,   Third Grade Student from Zhilin High School in
                   Zhejiang Provicne; 42 points, Gold Medal
Huang Jiajun,   First Grade Student from Shanghai Middle School in
                   Shanghai; 34 points, Gold Medal
Yuan Zhizhen,   Second Grade Student from Wugang No.3 Middle
                   School in Hubei Province; 42 points, Gold Medal
Yu Ranfeng,   Second Grade Student from Middle School Affiliated to
                   Nanjing Normal University; 35 points, Gold Medal

The top ten teams and their total scores are as follows:
1. China;   227 points
2. United States;   227 points
3. South Korea;   226 points
4. North Korea;   187 points
5. Thailand;   185 points
6. Russia;   179 points
7. Vietnam;   177 points
8. Singapore;   174 points
9. Serbia;   171 points
10. Poland   168 points

The gold medal line is 31 points, the silver medal line is 24 points, and the bronze medal line is 17 points.

The Six contestants of the Chinese National Team received the fist stage of training at the High School Affiliated to Nanjing Normal University, Jiangsu Province from May 4 to 12. Xiong Bin, Qu Zhenhua, Yu Hongbing, Shan Zun, Chen Yonggao, Lin Tianqi, and Lu Sheng gave reports and counseling to the students. Principal Ge Jun of the High School Affiliated to Nanjing Normal University has done a lot of work for the national team training.

From June 15th to 26th, the national team conducted the second phase of training in Shanghai Middle School. Xiong Bin, Qu Zhenhua, Yu Hongbing, Leng Gangsong, Yao Yijun, Fu Yunhao, Wang Bin, Zhang Sihui, Lin Bo, Chen Xiaomin gave special lectures and guidance to students. The principal of Shanghai Middle School Feng Zhigang and teacher Wang Guangting provided excellent logistical support for the training of the national team and did a lot of work.

From July 9th to 14th, the national team conducted pre-departure training and preparations at the School of Mathematical Sciences, Peking

University. Academician Yuan Yaxiang, Chairman of the Chinese Mathematical Society, Professor Chen Min, Vice Chairman of the Chinese Mathematical Society, Professor Wu Jianping, Deputy Director of the Chinese Mathematical Society Popularization Working Committee and Mathematical Olympiad Committee, Professor Hu Jun, Dean and Secretary of the School of Mathematical Sciences, Peking University, Professor Liu Bin, the former deputy dean, and Professor Li Ruo, the deputy dean, had a discussion with the students. Professor Sun Zhaojun, deputy secretary of the Party Committee of the School of Mathematical Sciences, Peking University, has done a lot of work for the national team. Professor Qiu Zonghu, former vice chairman of the Olympic Committee of the Chinese Mathematical Society, received the students and gave them pre-match guidance.

First Day
(8:30 – 13:00; July 16, 2019)

1 Let $\mathbb{Z}$ be the set of integers. Determine all functions $f : \mathbb{Z} \to \mathbb{Z}$ such that, for all integers a and b,

$$f(2a) + 2f(b) = f(f(a+b)).$$

(Contributed by South Africa)

Solution We denote the given equation (for specific a, b) as $P(a, b)$. Then from $P(0, x)$ we have

$$f(0) + 2f(x) = f(f(x)). \qquad ①$$

From $P(x, 0)$ we have

$$f(2x) + 2f(0) = f(f(x)). \qquad ②$$

Comparing ① and ②, we obtain

$$f(2x) = 2f(x) - f(0). \qquad ③$$

Plugging ③ into the original equation, we have

$$2f(a) - f(0) + 2f(b) = f(0) + 2f(a+b). \qquad ④$$

Let $g(x) = f(x) - f(0)$, then $g(0) = 0$, and we can rewrite ④ as

$$g(a+b) = g(a) + g(b). \qquad ⑤$$

Then Cauchy's method yields that $g(n) = g(1)n$ for every integer n. Thus $f(x) = kx + c$, where $k, c \in \mathbb{Z}$. Plugging into the original equation, we get

$$2k(a+b) + 3c = k^2(a+b) + (k+1)c.$$

for all integers a, b. This happens if and only if $2k = k^2$ and $3c = (k+1)c$. Thus, $k = 0$ or $k = 2$. If $k = 2$, then c can be any integer, while if $k = 0$, then $c = 0$.

In summary, the desired functions are $f(x) = 0$ and $f(x) = 2x + c$ (where c is an integer). □

2 In triangle ABC, point A_1 lies on side BC and point B_1 lies on side AC. Let P and Q be points on segments AA_1 and BB_1, respectively, such that PQ is parallel to AB. Let P_1 be a point on line PB_1, such that B_1 lies strictly between P and P_1, and $\angle PP_1C = \angle BAC$. Similarly, let Q_1 be a point on line QA_1, such that A_1 lies strictly between Q and Q_1, and $\angle CQ_1Q = \angle CBA$. Prove that points P, Q, P_1 and Q_1 are concyclic. (Contributed by Ukraine)

Solution   As shown in the figure 2.1, let rays AA_1, BB_1 intersect the circumcircle of $\triangle ABC$ at A_2, B_2, respectively. Since $PQ \parallel AB$, we have $\angle A_2B_2B = \angle A_2AB = \angle A_2PQ$, so A_2, B_2, P, Q are concyclic. Also since

$$\angle CA_2A = \angle CBA = \angle CQ_1Q = \angle CQ_1A_1$$

points C, A_1, A_2, Q_1 are concyclic. Hence,

$$\angle A_2Q_1Q = \angle A_2Q_1A_1 = \angle A_2CA_1 = \angle A_2CB = \angle A_2AB = \angle A_2PQ.$$

This implies that Q_1, A_2, Q, P are concyclic. Similarly, we can prove that P_1, B_2, P, Q are concyclic. Therefore, P_1, Q_1, P, Q, A_2, B_2 all lie on the same circle, and the desired conclusion follows. □

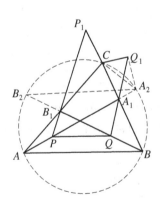

Fig. 2.1

3 A social network has 2019 users, some pairs of whom are friends. Whenever user A is friends with user B, user B is also friends with user A. Events of the following kind may happen repeatedly, one at a time:

Three users A, B, and C such that A is friends with both B and C, but B and C are not friends, change their friendship statuses such that B and C are now friends, but A is no longer friends with B, and no longer friends with C. All other friendship statuses are unchanged.

Initially, 1010 users have 1009 friends each, and 1009 users have 1010 friends each. Prove that there exists a sequence of such events after which each user is friends with at most one other user. (Contributed by Croatia)

Solution (by Deng Mingyang) We interpret this problem in the language of graph theory. First, there is a simple graph with 2019 vertices, in which 1010 vertices have degree 1009, and the other 1009 vertices have degree 1010. The following operation is allowed: if A is adjacent to B and C, while B, C are not adjacent, then we delete edges AB, AC and connect BC. We proceed to show that there is a sequence of operations such that after these operations, the graph has only isolated vertices and edges with no common vertices.

Note that for any two vertices u, v in G, the sum of their degrees is at least 2018, so it follows that either u, v are adjacent to each other, or they have a common neighbor. Hence G is connected. Apparently, G is not a complete graph, and it contains vertices with odd degrees. In fact, we shall show that the desired conclusion holds for every simple graph that is connected, non-complete, and contains a vertex with odd degree. In the following argument we only assume that G has these properties.

Step 1: If G is not a tree, then we claim that there exists a sequence of operations, after which G becomes a tree. Note that an operation does not change the parity of the degree of any vertex (if a vertex has odd degree, its degree remains odd after any sequence of operations, and vice versa). Thus, after any sequence of operations G still has a vertex with odd degree. It suffices to show that if G is not a tree, then there exists an operation that leaves G connected, so that since the number of edges decreases strictly, it will eventually become a tree after finitely many operations.

We choose the longest cycle in G, namely $x_1, x_2, \ldots, x_k, x_1$. If it is not a Hamiltonion cycle, then there is a vertex (which we assume to be x_1) that is adjacent to some vertex y outside the cycle. Then y is not adjacent to x_2, since otherwise $x_1, y, x_2, \ldots, x_k, x_1$ is a longer cycle. Then we delete edges $x_1 y, x_1 x_2$ and add $y x_2$. It is easy to see that the graph remains connected.

On the other hand, if the cycle is Hamiltonion, since there is a vertex with odd degree and the graph is not complete, the graph must contain an edge, which we assume to be $x_1 x_i$, such that either x_1 is not adjacent to x_{i+1}, or x_1 is not adjacent to x_{i-1}. Without loss of generality we assume x_1 is not adjacent to x_{i+1}. Then we may delete edges $x_1 x_i, x_i x_{i+1}$ and add $x_1 x_{i+1}$, and the graph remains connected.

Step 2: If G is a tree, we claim that we may do a sequence of operations to change it into a graph with only isolated vertices and edges with no common vertices. If G has only 1 or 2 vertices, then the conclusion holds automatically. Suppose G has at least 3 vertices, then there exists a vertex u whose degree is at least 2. Let uv, uw be two edges incident to u, then v is not adjacent to w. Now we delete edges uv, uw, and add vw. Then the graph is divided into two connected components, each of which is still a tree. For every component with at least 3 vertices we continue with the above operation, until every component has at most 2 vertices. Finally we obtain the desired configuration. $\square$

Second Day
(8:30 – 13:00; July 17, 2019)

4 Find all pairs (k, n) of positive integers such that
$$k! = (2^n - 1)(2^n - 2)(2^n - 4) \cdots (2^n - 2^{n-1}).$$
(Contributed by El Salvador)

Solution Denote the expression on the right side as R_n, then
$$v_2(R_n) = \sum_{i=0}^{n-1} v_2(2^n - 2^i) = \sum_{i=0}^{n-1} i = \frac{1}{2} n(n-1).$$
It is well known that $v_2(k!) = k - S_2(k)$, where $S_2(k)$ denotes the sum of digits in the decimal representation of k. Hence
$$k - S_2(k) = \frac{1}{2} n(n-1),$$
and $k = \frac{1}{2} n(n-1) + S_2(k) \geq \frac{1}{2} n(n-1) + 1.$

Checking $n = 1, 2, 3, 4$ directly, we get two solutions $(k, n) = (1, 1), (3, 2)$. For $n \geq 5$, we claim that

$$\left(\frac{1}{2}n(n-1)+1\right)! > 2^{n^2}, \tag{1}$$

while $R_n \leq (2^n)^n = 2^{n^2}$, so the equation has no solution.

For $n = 5$, we directly verify that $11! = 39916800 > 2^{25} = 33554432$. For $n > 5$, we have

$$\left(\frac{1}{2}n(n-1)+1\right)! = 11! \cdot \prod_{i=12}^{\frac{1}{2}n(n-1)+1} i > 2^{25} \cdot 8^{\frac{1}{2}n(n-1)-10}$$

$$= 2^{25+\frac{3}{2}n(n-1)-30} \geq 2^{\frac{3}{2}n(n-1)-5} \geq 2^{n^2}.$$

In summary, $(k, n) = (1, 1), (3, 2)$ are the only pairs that satisfy the equation. □

5 The Bank of Bath issues coins with an H on one side and a T on the other. Harry has n of these coins arranged in a line from left to right. He repeatedly performs the following operation: if there are exactly $k > 0$ coins showing H, then he turns over the kth coin from the left; otherwise, all coins show T and he stops. For example, if $n = 3$ the process starting with the configuration THT would be $THT \to HHT \to HTT \to TTT$, which stops after three operations.

(a) Show that, for each initial configuration, Harry stops after a finite number of operations.

(b) For each initial configuration C, let $L(C)$ be the number of operations before Harry stops. For example, $L(THT) = 3$ and $L(TTT) = 0$. Determine the average value of $L(C)$ over all 2^n possible initial configurations C. (Contributed by United States)

Solution 1 Let $V_n = \{H, T\}^n$ be the set of all sequences of H and T with length n, i.e., the set of all statuses of the n coins. We apply induction to show that for any $C \in V_n$, $L(C)$ is finite, and

$$\sum_{C \in V_n} L(C) = 2^{n-2}n(n+1).$$

For $n = 1$, we have $V_1 = \{H, T\}, L(H) = 1, L(T) = 0$. For $n = 2$, $V_2 = \{HH, HT, TH, TT\}, L(HH) = 2, L(HT) = 1, L(TH) = 3, L(TT) = 0$. Thus the assertion holds when $n = 1, 2$.

Suppose $n \geq 3$, and the assertion already holds for all smaller n. For two sequences A and B, let $A \cdot B$ denote the sequence obtained by concatenating the two sequences together (B is placed behind A). Let

$$X = \{C \cdot T \mid C \in V_{n-1}\}, \qquad Y = \{H \cdot C \mid C \in V_{n-1}\},$$
$$Z = \{T \cdot C \cdot H \mid C \in V_{n-2}\}, \quad W = \{H \cdot C \cdot T \mid C \in V_{n-2}\}.$$

Apparently $V_n = X \cup Y \cup Z$, and $X \cap Y = W$.

For $C \cdot T \in X$, since the last term is T, operations on $C \cdot T$ are effectively the same as operations on C, so it becomes the sequence $TT \cdots T$ after $L(C)$ operations, i.e., $L(C \cdot T) = L(C)$.

For $H \cdot C \in Y$, suppose H appears k times in C, then H appears $k+1$ times in $H \cdot C$, and the $k+1^{\text{th}}$ term in $H \cdot C$ is the same as the kth term in C. Hence, the operations on $H \cdot C$ are effectively first operating on C, to obtain $HTT \cdots T$ after $L(C)$ operations, and then operating on the first term to obtain a sequence of all T. Therefore, $L(H \cdot C) = L(C) + 1$.

For $T \cdot C \cdot H \in Z$, suppose H appears k times in C, then H appears $k+1$ times in $T \cdot C \cdot H$, and the $k+1^{\text{th}}$ term in $T \cdot C \cdot H$ is the same as the kth term in C. Thus for $T \cdot C \cdot H$, we first operate on C, obtaining $TT \cdots TH$ after $L(C)$ operations, then change every T into H, from left to right, and finally change every H into T, from right to left. Hence, the process ends after $L(T \cdot C \cdot H) = L(C) + 2n - 1$ operations.

For $H \cdot C \cdot T \in W$, from the discussion in the previous cases we have

$$L(H \cdot C \cdot T) = L(H \cdot C) = L(C) + 1.$$

So far, we have shown that for every $C \in V_n$, $L(C)$ is a finite number. Finally, apply the induction hypothesis and we have

$$\sum_{C \in V_n} L(C)$$

$$= \sum_{C \cdot T \in X} L(C \cdot T) + \sum_{H \cdot C \in Y} L(H \cdot C) + \sum_{T \cdot C \cdot H \in Z} L(T \cdot C \cdot H)$$
$$- \sum_{H \cdot C \cdot T \in W} L(H \cdot C \cdot T)$$

$$= \sum_{C \in V_{n-1}} L(C) + \sum_{C \in V_{n-1}} (L(C) + 1) + \sum_{C \in V_{n-2}} (L(C) + 2n - 1)$$
$$- \sum_{C \in V_{n-2}} (L(C) + 1)$$

$$= 2 \sum_{C \in V_{n-1}} L(C) + 2^{n-1} + (2n-2)2^{n-2} = 2 \cdot 2^{n-3} n(n-1) + 2n \cdot 2^{n-2}$$

$$= 2^{n-2} n(n+1).$$

Therefore, the average of all $L(C)$ is $\dfrac{1}{4} n(n+1)$.

Solution 2   If every term of C is T, then $L(C) = 0$. Suppose C contains i instances of H $(0 < i \le n)$, and we assume the positions of H are $0 < a_1 < a_2 < \cdots < a_i \le n$. Let $a_0 = 0$, then since $a_i \ge i$, there exists a unique integer j with $0 < j \le i$ and $a_{j-1} < i \le a_j$. From the rule of operation we see that the $1, 2, \ldots, (a_j - i)$-th operations change the T on positions $i, i+1, \ldots, a_j - 1$ into H, respectively. Then the $(a_j - i + 1)^{\text{th}}$ operation changes the H on position a_j into T, and the next $a_j - i$ operations change the H on positions $a_j - 1$ through i into T, in the reverse order. Hence, after $2a_j - 2i + 1$ operations the overall effect is changing the H on position a_j into T, and leaving all other terms unchanged. These $2a_j - 2i + 1$ operations together compose one group of operations, and repeating such process, we finally change all the H into T after i groups of operations. Therefore the assertion in (a) is true.

Suppose the result of the kth group of operations is changing the H on position $a_{j(k)}$ into T, and before the kth group there are $i + 1 - k$ instances of H, then the kth group contains exactly $2a_{j(k)} - 2(i+1-k) + 1$ operations. Since $j(1), j(2), \ldots, j(i)$ form a permutation of $1, 2, \ldots, i$, it follows that

$$L(C) = \sum_{k=1}^{i} (2a_{j(k)} - 2(i+1-k) + 1) = 2 \left(\sum_{k=1}^{i} a_k \right) - i^2.$$

If $A = \{a_1, a_2, \cdots, a_i\} \subset \{1, 2, \cdots, n\}$, let $S(A)$ denote the sum of elements in A, then

$$L(C) = 2S(A) - |A|^2.$$

The same result holds when $A = \varnothing$. Let $\bar{A} = \{1, 2, \ldots, n\} \backslash A$, then $S(A) + S(\bar{A}) = \dfrac{1}{2} n(n+1)$ and $|A| + |\bar{A}| = n$. Pairing up A and $\bar{A}$, we may calculate that

$$\sum_{A} S(A) = 2^{n-2} n(n+1), \quad \sum_{A} |A| = 2^{n-1} n.$$

Hence,

$$\sum_C L(C) = \sum_A 2S(A) - \sum_A |A|^2 = 2^{n-1}n(n+1) - \sum_A |A|$$

$$- \sum_A |A|(|A|-1) = 2^{n-1}n(n+1) - 2^{n-1}n - \sum_{i=0}^{n} i(i-1)\binom{n}{i}$$

$$= 2^{n-1}n^2 - \sum_{k=0}^{n-2} n(n-1)\binom{n-2}{k}$$

$$= 2^{n-1}n^2 - 2^{n-2}n(n-1) = 2^{n-2}n(n+1).$$

Therefore, the average of all $L(C)$ is $\dfrac{1}{4}n(n+1)$. $\Box$

6 Let I be the incentre of acute triangle ABC with $AB \neq AC$. The incircle ω of ABC is tangent to sides BC, CA and AB at D, E and F, respectively. The line through D perpendicular to EF meets ω again at R. Line AR meets ω again at P. The circumcircles of triangles PCE and PBF meet again at Q. Prove that lines DI and PQ meet on the line through A perpendicular to AI. (Contributed by India)

Solution 1   As shown in Fig. 6.1, let the exterior angle bisector of $\angle BAC$ and DI intersect at L, then $AL \perp AI$. It suffices to show that L, Q, P are

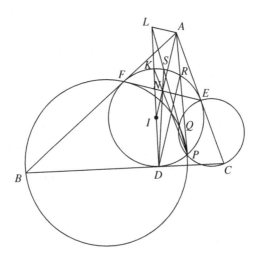

Fig. 6.1

collinear. Let K be the second intersection of DL and ω, and let N be the midpoint of EF. We proceed with the followings:

(i) K, N, P are collinear. Since $RFPE$ is a harmonic quadrilateral, we see that EF bisects $\angle RNP$, and since

$$\frac{1}{2}\widehat{KF} + \frac{1}{2}\widehat{FD} = 90° = \frac{1}{2}\widehat{RE} + \frac{1}{2}\widehat{FD},$$

we have $\widehat{KF} = \widehat{RE}$, so K, R are symmetric with respect to AI. Hence $\angle KNF = \angle RNE = \angle PNE$, so K, N, P are collinear.

(ii) L, S, P are collinear. Since A is the image of N under inversion with respect to ω, and $LA \perp AI$, we see that AL is the polar of N with respect to ω. Let PS and DK intersect at L', then the polar of L' passes through N, and the polar of N passes through L. Hence L' lies on the line AL, and $L = L'$. This shows that L, S, P are collinear.

Now it suffices to show that S, Q, P are collinear, as shown in Fig. 6.2. We apply the notation of directed angles, in order to avoid dependence on specific configuration of the points. Let $\angle(a, b)$ denote the smallest non-negative angle α, such that if we rotate the line a counterclockwise by α, we will obtain a line that is parallel to b. In the following argument, the arithmetic of directed angles is considered modulo π. Since B, F, Q, P are

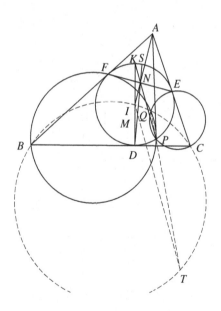

Fig. 6.2

concyclic and C, E, Q, P are concyclic, we have

$$\angle(BQ, QC) = \angle(BQ, QP) + \angle(PQ, QC) = \angle(BF, FP) + \angle(PE, EC)$$
$$= \angle(EF, EP) + \angle(FP, FE) = \angle(FP, EP) = \angle(DF, DE)$$
$$= \angle(BI, IC).$$

Hence, B, I, Q, C are concyclic. Let T be the second intersection of QP and $\odot BIQC$, and IT intersects DS at M.

(iii) M is the midpoint of DN. Note that

$$\angle(BI, IT) = \angle(BQ, QT) = \angle(BF, FP) = \angle(FK, KP),$$

and $FD \perp FK$, $FD \perp BI$, we have $FK \parallel BI$, so that $IT \parallel KNP$. Since I is the midpoint of DK, M is also the midpoint of DN.

Finally we show that S, P, T are collinear, hence completing the proof. As shown in Fig. 6.3, let F_1, E_1 be the midpoints of F_1, E_1. Since

$$DF_1 \cdot F_1 F = DF_1^2 = BF_1 \cdot F_1 I,$$

we see that F_1 lies on the radical axis of ω and $\odot BIC$. Similarly E_1 lies on the radical axis of ω and $\odot BIC$, so the line $E_1 F_1$ is exactly the radical

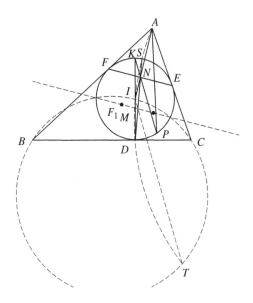

Fig. 6.3

axis of ω and $\odot BIC$. Since M is the midpoint of DN, M lies on $E_1 F_1$, so M has the same power to circles ω and $\odot BIC$. Therefore,

$$DM \cdot MS = IM \cdot MT,$$

and S, I, D, T are concyclic. Thus,

$$\angle(DS, ST) = \angle(DI, IT) = \angle(DK, KP) = \angle(DS, SP),$$

which is the same as saying S, P, T are collinear.

Solution 2 (By Huang Jiajun) As shown in Fig. 6.4, let Ω be the circumcircle of $\triangle ABC$, and the midpoints of arcs $\overparen{BAC}$ and $\overparen{BC}$ are M_a, N_a, respectively. Let L be the intersection of the lines DI and AM_a. Note that since $AM_a \perp AN_a$ and A, I, N_a are collinear, we have $AL \perp AI$. Then it suffices to show that P, Q, L are collinear.

We construct circle ω with N_a as its center and IN_a as its radius. Then by the property of the incenter, B, C both lie on ω. Suppose line $M_a I$ intersects Ω, ω again at T, P', respectively.

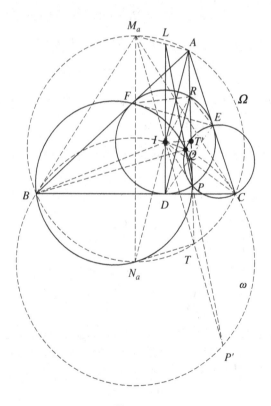

Fig. 6.4

(i) The concave quadrilaterals $AERF$ and M_aCIB are similar. First we have $AE = AF$ and $M_aC = M_aB$, and

$$\angle(FA, AE) = \angle(BA, AC) = \angle(BM_a, M_aC),$$

so $\triangle EAF \sim \triangle CM_aB$. Also

$$\angle(RE, EF) = \frac{\pi}{2} - \angle(DR, RE) = \frac{\pi}{2} - \angle(DE, EC) = \angle(IC, CB),$$

and similarly, $\angle(EF, FE) = \angle(IB, BC)$. Therefore, $AERF \sim M_aCIB$.
(ii) A, P, T are collinear.

It follows by (i) that in similar quadrilaterals $AERF$ and M_aCIB, $\odot I$ and ω are corresponding circles, and AR, M_aI are corresponding lines, so P, P' are corresponding points. Hence,

$$\angle(PA, AE) = \angle(P'M_a, M_aC) = \angle(TM_a, M_aC) = \angle(TA, AC).$$

Thus, A, P, T are collinear.
(iii) P, Q, P' are collinear. Since B, P, Q, F are concyclic, we have

$$\angle(BQ, QP) = \angle(BF, FP) = \angle(FR, RP),$$

and similarly $\angle(PQ, QC) = \angle(PR, RE)$. Hence,

$$\angle(BQ, QC) = \angle(BQ, QP) + \angle(PQ, QC) = \angle(FR, RP) + \angle(PR, RE)$$
$$= \angle(FR, RE) = \angle(BI, IC).$$

Then it follows that B, I, Q, C are concyclic, or equivalently Q lies on ω. Consequently,

$$\angle(BQ, QP') = \angle(BI, IP') = \angle(BI, IM_a) = \angle(FR, RA) = \angle(FR, RP)$$
$$= \angle(BF, FP) = \angle(BQ, QP),$$

and therefore P, Q, P' are collinear.
(iv) P, Q, L are collinear. By Menelaus' theorem it suffices to show that

$$\frac{AP}{PT} \cdot \frac{TP'}{P'M_a} \cdot \frac{M_aL}{LA} = 1.$$

Construct the perpendicular line of PR through I, and let T' be its foot. Note that since $N_aT \perp IP'$, T and T' are corresponding points in quadrilaterals $AERF$ and M_aCIB. This implies

$$\frac{TP'}{P'M_a} = \frac{T'P}{PA}.$$

In right triangles ITT' and $M_a AN_a$ we have

$$\angle(T'T, TI) = \angle(AT, TM_a) = \angle(AN_a, N_a M_a),$$

so $\triangle ITT' \sim \triangle M_a N_a A$. Combining with $AERF \sim M_a CIB$, we obtain that

$$\angle(IP, PT) = \angle(AR, RI) = \angle(M_a I, IN_a).$$

Thus, P, I are corresponding points in similar triangles ITT' and $M_a N_a A$. Combining with $IL \parallel M_a N_a$, we have

$$\frac{M_a L}{LA} = \frac{N_a I}{IA} = \frac{TP}{PT'}.$$

Thus,

$$\frac{AP}{PT} \cdot \frac{TP'}{P'M_a} \cdot \frac{M_a L}{LA} = \frac{AP}{PT} \cdot \frac{T'P}{PA} \cdot \frac{TP}{PT'} = 1,$$

which yields that P, Q, L are collinear.

Combining the above results, we obtain the desired conclusion. $\square$

Solution 3 (By Xie Baiting) As shown in Fig. 6.5, on the circumcircle of $\triangle ABC$, let N and L be the midpoints of $\overset{\frown}{BC}$ and $\overset{\frown}{BAC}$, respectively. Construct circle $\odot N$ with radius NI, so that it passes through B, C. Let LI intersect $\odot N$ again at U, and intersect the circumcircle of $\triangle ABC$ again at S. Suppose the circumcircle of $\triangle AFE$ (denoted as Ω_1) and the circumcircle of $\triangle ABC$ (denoted as Ω) intersect at A, M. Let V be the intersection of DI, AL. We proceed to prove the following assertions:

(i) A, R, S are collinear.

Let $\angle BAC = 2\alpha, \angle ABC = 2\beta, \angle ACB = 2\gamma$. Note that $\angle BSI = \angle ISC$, and

$$\angle IBS = \angle IBC + \angle SBC = \beta + \angle CLI = \gamma + (\beta - \gamma) + \angle CLI$$

$$= \angle ICA + \angle LCA + \angle CLI = \angle CIS,$$

so we have $\triangle IBS \sim \triangle CIS$. Hence,

$$\frac{\sin \angle BAS}{\sin \angle CAS} = \frac{BS}{CS} = \frac{BS}{IS} \cdot \frac{IS}{CS} = \left(\frac{BI}{IC}\right)^2 = \left(\frac{\sin \gamma}{\sin \beta}\right)^2.$$

Since I is the center of ω, $DR \perp EF$, we have $\angle EDR = \angle FDI = \beta$, $\angle FDR = \angle EDI = \gamma$. Thus,

$$\frac{\sin \angle BAR}{\sin \angle CAR} = \frac{\sin \angle AFR}{\sin \angle RFE} \cdot \frac{\sin \angle REF}{\sin \angle REA} = \left(\frac{\sin \angle FDR}{\sin \angle EDR}\right)^2 = \left(\frac{\sin \gamma}{\sin \beta}\right)^2$$

$$= \frac{\sin \angle BAS}{\sin \angle CAS}.$$

Therefore A, R, S are collinear, and assertion (i) is proven.

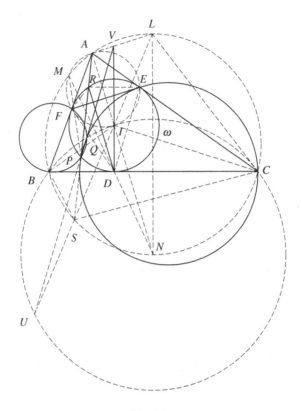

Fig. 6.5

(ii) U, P, Q are collinear.

Since $\angle AFE = \angle AEF = \angle LBC = \angle LCB = 90° - \alpha$, it follows that $\triangle AFE \sim \triangle LBC$ (rotationally), and M is the rotation canter.

Now AI, LN are diameters of Ω_1 and Ω, respectively, so I, N are corresponding points in the pair of similar triangles, and thus ω and $\odot N$ are corresponding circles. Also $\angle FAR = \angle BAS = \angle BLS$, so AR, LS are corresponding lines, so the intersection points of AR and ω, namely R, P, should correspond to the intersection points of LS and $\odot N$, namely I, U.

Hence, $\angle BIU = \angle FRP = \angle BFP = \angle BQP$, and from

$$\angle BQC = \angle BAC + \angle ABQ + \angle ACQ = \angle BAC + \angle FPQ + \angle EPQ$$

$$= \angle BAC + \angle FPE = 2\alpha + \angle FDE = 90° + \alpha = \angle BIC,$$

we see that Q lies on $\odot N$. Therefore, $\angle BQU = \angle BIU = \angle BQP$, or equivalently U, P, Q are collinear, and (ii) is proven.

(iii) A, M, P, V are concyclic.

From the rotational similarity of $\triangle AFE$ and $\triangle LBC$ in (ii) (where M is the rotation center), we obtain that $\triangle MFB \sim \triangle MEC$, so that $\dfrac{MB}{MC} = \dfrac{FB}{EC} = \dfrac{BD}{CD}$, and MD bisects $\angle BMC$. Thus, M, D, N are collinear.

Next, $DV \parallel LN$ implies that $\angle MDV = \angle MNL = 180° - \angle MAV$, so A, M, D, V are concyclic. Since $RD \perp EF, AN \perp EF$, we have $RD \parallel AN$, so $\angle PAN = \angle PRD = \angle PDB$. Also

$$\angle MAN = \angle MAB + \alpha = \angle MCB + \angle CMN = \angle MDB,$$

so $\angle MAP = \angle MAN - \angle PAN = \angle MDB - \angle PDB = \angle MDP$, and A, M, P, D are concyclic.

Thus, A, M, P, D, V are concyclic, and (iii) is proven.

(iv) U, V, P, Q are concyclic.

By (iii) we have $\angle MPV = 180° - \angle MAV = 90° - \angle MAN = \angle MIA = \angle MFA$.

In the rotational similarity in (ii), the points M, F, P correspond to M, B, U, respectively. Hence,

$$\angle MPU = \angle MFB = 180° - \angle MFA = 180° - \angle MPV,$$

which means U, V, P are collinear. Combining with (ii), we conclude that U, V, P, Q are collinear, and (iv) is proven.

In combination, we derive the conclusion in the original problem. □

Solution 4 (By Yuan Zhizhen) As shown in Fig. 6.6, let L be the intersection of DI and the exterior angle bisector of $\angle BAC$, then $AL \perp AI$. It suffices to show that L, P, Q are collinear.

Consider the inversion with respect to ω. Let the images of $A, B, C, \cdots$ be $A', B', C', \cdots$, respectively. Then the points on ω are mapped to themselves, and B', C' are the midpoints of DF, DE, respectively. Let N be the midpoint of EF, and construct the line through N perpendicular to DI, with L' as its foot. In order to prove that L, Q, P are collinear, it suffices to show that I, L', P', Q' are concyclic.

Let K be the second intersection of DI and ω, and let $EF, B'C'$ intersect DK at U, V, respectively. Let X be the intersection of the lines $B'C', PK$, and let Y, Z be the midpoints of KU, KN, respectively.

(i) P, N, K are collinear.

Since $KR \perp DR, DR \perp EF$, we have $KR \parallel EF$, so $\widehat{FK} = \widehat{ER}$, and $\angle FRK = \angle EPR$. Note that $ERFP$ is a harmonic quadrilateral, so by the property of harmonic quadrilaterals, N lies on PK.

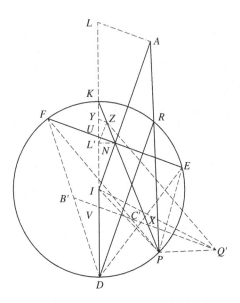

Fig. 6.6

(ii) Q', B', C' are collinear.

Since P, B, F, Q are concyclic and P, C, E, Q are concyclic, under inversion we obtain that P, B', F, Q' are concyclic and P, C', E, Q' are concyclic. Hence,

$$\angle(B'Q', Q'P) = \angle(B'F, FP) = \angle(DF, FP) = \angle(DE, EP)$$
$$= \angle(C'E, EP) = \angle(C'Q', Q'P).$$

Therefore, Q', B', C' are collinear.

(iii) P, Q', K, V are concyclic and $\triangle PQ'X \sim \triangle VKX$.

Since P, B', F, Q' are concyclic, we have

$$\angle(VQ', Q'P) = \angle(B'Q', Q'P) = \angle(B'F, FP)$$
$$= \angle(DF, FP) = \angle(DK, KP) = \angle(VK, KP),$$

which implies P, Q', K, V are concyclic, and thus $\triangle PQ'X \sim \triangle VKX$.

(iv) $\triangle PIQ' \sim \triangle XZQ'$ and P, I, Z, Q' are concyclic.

Note that

$$\angle(PK, PI) = \angle(IK, KP) = \angle(DK, KP) = \angle(DE, EP)$$
$$= \angle(C'E, EP) = \angle(C'Q', Q'P),$$

so we have

$$\angle(Q'P,\ PI) = \angle(Q'P, PK) + \angle(PK, PI)$$
$$= \angle(Q'P, PK) + \angle(C'Q',\ Q'P) = \angle(C'Q',\ PK)$$
$$= \angle(Q'X,\ XZ).$$

Since V is the midpoint of DU, we have

$$\overrightarrow{VY} = \overrightarrow{VU} + \overrightarrow{UY} = \frac{1}{2}\overrightarrow{DU} + \frac{1}{2}\overrightarrow{UK} = \frac{1}{2}\overrightarrow{DK},$$

so $VY = \dfrac{1}{2}DK = IP$. Combining with $YZ \parallel UN \parallel VX$ and $\triangle PQ'X \sim VKX$, we obtain

$$\frac{IP}{ZX} = \frac{VY}{ZX} = \frac{KV}{KX} = \frac{Q'P}{Q'X}.$$

Hence, $\triangle PIQ' \sim \triangle XZQ'$, and

$$\angle(PI, IQ') = \angle(XZ, ZQ') = \angle(PZ, ZQ').$$

Therefore, P, I, Z, Q', are concyclic.

(v) P, I, L', Z are concyclic.

Since $KL' \perp L'N$ and $KZ = ZN$, we have

$$\angle(L'Z, ZP) = \angle(L'Z, ZN) = 2\angle(L'K, KN) = 2\angle(DK, KP) = \angle(L'I, IP).$$

Hence, P, I, L', Z are concyclic.

Combining (iv) and (v), we obtain that I, L', Z, Q', P are concyclic, and by the property of inversion we conclude that L, P, Q, hence completing the proof. $\qquad\square$

Solution 5 (By Hu Sulin) As shown in Fig. 6.7, let L be the intersection of DI and the exterior angle bisector of $\angle BAC$, then $AL \perp AI$. It suffices to show that L, P, Q are collinear.

Consider the inversion with respect to ω. Let the images of $A, B, C, \ldots$ be $A', B', C', \ldots$, respectively. Then the points on ω are mapped to themselves, and B', C' are the midpoints of DF, DE, respectively. Let N be the midpoint of EF, and construct the line through N perpendicular to DI, with L' as its foot. In order to prove that L, Q, P are collinear, it suffices to show that I, L', P', Q' are concyclic.

Let K be the second intersection of DI and ω. Denote the circumcircle of $\triangle PB'F$ as Γ_1, and denote the circumcircle of $\triangle PC'E$ as Γ_2. It follows from the property of inversion that Q' is the second intersection of Γ_1 and

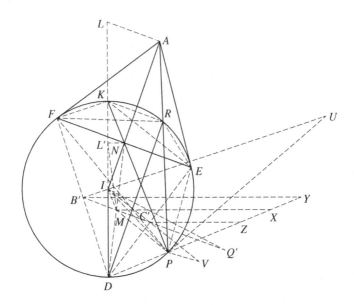

Fig. 6.7

Γ_2 (apart from P). Denote the circumcircle of $\triangle IPL'$ as γ. Thus It suffices to show that $\Gamma_1, \Gamma_2, \gamma$ are coaxial.

Lemma (Casey's circle-power theorem) *Let Γ_1, Γ_2 be two circles on the plane. Let $p(T, \Gamma_i)$ denote the power of T to the circle $\Gamma_i (i = 1, 2)$. For a real number $\lambda \neq 0, 1$, the set of points T such that $\dfrac{p(T, \Gamma_1)}{p(T, \Gamma_2)} = \lambda$ is a circle that is coaxial with Γ_1, Γ_2.*

Proof of lemma   Let the standard equation of Γ_1, Γ_2 (i.e., the equation in the form $x^2 + y^2 + dx + ey + f = 0$) be $F_1(x, y) = 0$ and $F_2(x, y) = 0$, respectively. Then the locus of P is determined by the equation $F_1 - \lambda F_2 = 0$. Since $\lambda \neq 1$, this equation gives a circle, whose standard equation is $\dfrac{F_1 - \lambda F_2}{1 - \lambda} = 0$. The radical axis of this circle and Γ_1 is then

$$\frac{F_1 - \lambda F_2}{1 - \lambda} = F_1 \Longleftrightarrow F_1 - F_2 = 0.$$

Hence, this line is the same as the radical axis of Γ_1 and Γ_2, which proves the lemma.

     Returning to the original problem, let M be the intersection of $B'C', DN$, and the circles $\Gamma_1, \Gamma_2, \gamma$ intersect the line DP at Y, Z, X

(different from P), respectively. Suppose FU, EV are diameters of Γ_1, Γ_2, respectively.

(i) P, N, K are collinear.

Since $\triangle AER \sim \triangle APE, \triangle AFR \sim APF$, we have

$$\frac{ER}{PF} = \frac{AE}{AP} = \frac{AF}{AP} = \frac{FR}{PF'},$$

so that $ER \cdot PF = FR \cdot PE$. Combining with $DR \perp EF$, we obtain

$$\angle(KD, DF) = \frac{\pi}{2} - \angle(FE, ED) = \angle(ED, DR).$$

Hence, $ER = FK$, $FR = EK$, and $FK \cdot PF = EK \cdot PE$. Combining with $\angle(PF, FK) = \angle(PE, ER)$, we have $S_{\triangle PFK} = S_{\triangle PEK}$, which implies that P, N, K are collinear.

(ii) M lies on γ and $MX \perp DK$.

Note that M is the midpoint of DN and $DL' \perp NL'$, so

$$\angle(IL', L'M) = \angle(DM, DI).$$

Also since $IM \parallel KN$ and $KP \perp DP$, we have $IM \perp DP$, so

$$\angle(DM, DI) = (\angle IP, PM).$$

Thus, $\angle(IL', L'M) = \angle(IP, PM)$, which means I, M, P, L' are concyclic, or equivalently M lies on γ. Further,

$$\angle(MX, XP) = \angle(MI, IP) = \angle(KP, PI) = \angle(DK, KP).$$

Since $XP \perp KP$, we obtain $MX \perp DK$.

(iii) $B'Y \perp DK, C'Z \perp DK$, and $\dfrac{p(X, \Gamma_1)}{p(X, \Gamma_2)} = -1$.

Since B', P, F, Y are concyclic, we have

$$\angle(B'Y, YP) = \angle(B'F, FP) = \angle(DF, FP) = \angle(DK, KP).$$

From $YP \perp KP$ we have $B'P \perp DK$, and similarly $C'Z \perp DK$. Combining with (ii) we see that $MX \parallel B'Y \parallel C'Z$. Since M is the midpoint of $B'C'$, it follows that X is the midpoint of YZ. Hence,

$$\frac{p(X, \Gamma_1)}{p(X, \Gamma_2)} = \frac{\overline{XP} \cdot \overline{XY}}{\overline{XP} \cdot \overline{XZ}} = -1.$$

(iv) $\dfrac{p(I, \Gamma_1)}{p(I, \Gamma_2)} = -1$.

Note that $FB' \perp B'U$ and $FB' \perp B'I$, so B', I, U are collinear. Similarly C', I, V are collinear. Combining with $VP \perp PE$ and $UP \perp PF$, we have

$$\angle(IU, UP) = \angle(DF, FP) = \angle(DE, EP) = \angle(IV, VP).$$

So that I, P, U, V are concyclic. Hence,

$$\frac{IU}{IV} = \frac{\sin \angle(UP, PI)}{\sin \angle(VP, PI)} = \frac{\cos \angle(PF, PI)}{\cos \angle(EP, PI)} = \frac{PF}{PE}.$$

Combining with the conclusion of (i) we obtain that

$$\frac{IU}{IV} = \frac{PF}{PE} = \frac{KE}{KF} = \frac{IC'}{IB'},$$

i.e., $IB' \cdot IU = IC' \cdot IV$. Therefore,

$$\frac{p(I, \Gamma_1)}{p(I, \Gamma_2)} = \frac{\overline{IB'} \cdot \overline{IU}}{\overline{IC'} \cdot \overline{IV}} = -1.$$

Results (iii) and (iv) together imply that $\gamma, \Gamma_1, \Gamma_2$ are coaxial. Since Q' lies on both Γ_1 and Γ_2, it also lies on γ, so I, L', P, Q' are concyclic. This completes the proof. $\qquad\square$

International Mathematical Olympiad

2020 (Saint Petersburg, Russia)

Affected by the new crown epidemic, the 61st International Mathematical Olympiad (IMO) originally scheduled to be held in St. Petersburg, Russia in July this year, has been postponed to September 18-28, 2020. It will be held online and remotely. From 105 countries and regions 616 students participated in the competition. In order to ensure the confidentiality and fairness of the competition, each participating country and region has set up an independent test center in accordance with the requirements of the organizing committee, and a neutral International Mathematical Olympiad commissioner will supervise the test on site. The test questions will be sent to the team leader interpreter 3 hours before the test. The test questions are sent to each team leader for translation 3 hours before the test, and the confirmed test questions are sent to the IMO commissioner 15 minutes before the test for printing, and then sent to the students. The test time is selected according to Greenwich Mean Time for each participating country and region. The time window is 4 hours and 30 minutes. The time window given by the organizing committee is 7:30 am to 12 noon GMT. All participating teams must choose to start the exam during this time period Time. The Chinese team chose the earliest time to start the exam, which is 7:30 am GMT (15:30 pm Beijing time).

After two exams on September 21st and 22nd, the Chinese team won the first place with a total score of 215 points.

The members of the Chinese team are as follows:

Team Leader: Xiong Bin, East China Normal University

Deputy Leader: He Yijie, East China Normal University

Observers: Yao Yijun, Fudan University
Qu Zhenhua, East China Normal University
Fu Yunhao, South Science and Technology University

Team Contestants:

Li Jinmin,	Second Grade Student from Bashu Middle School in Chongqing; 42 points, Gold Medal
Han Xinmiao,	Third Grade Student from Zhilin Middle School in Zhejiang Province; 40 points, Gold Medal
Yi Jia,	Third Grade Student from Middle School Affiliated to Renmin University of China; 37 points, Gold Medal
Liang Jingxun,	First Grade Student from Hangzhou City Xuejun Middle School in Zhejiang Province; 36 points, Gold Medal
Rao Rui,	Second Grade Student from Middle School Affilated to South Normal University; 31 points, Gold Medal
Yan Binwei,	Third Grade Student from Middle School Affiliated to Nanjing Normal University; 29 points, Silver Medal

The top ten teams and their total scores are as follows:

1. China;   215 points
2. Russia;   185 points
3. United States;   183 points
4. South Korea;   175 points
5. Thailand;   174 points
6. Italy;   171 points
7. Poland;   171 points
8. Australia;   168 points
9. United Kingdom; 167 points
10. Brazil;   164 points

From September 17th to 19th, the Chinese national team conducted training and preparation before the competition at the School of Mathematical Sciences of Peking University. Academician Tian Gang, Chairman of the Chinese Mathematical Society, Professor Chen Min, former Vice Chairman of the Chinese Mathematical Society, and Mathematics Science of Peking University Professor Chen Dayue, the dean of the college, Professor Sun Zhaojun, the deputy dean, and Professor Wu Jianping, Professor

Liu Bin, and Professor Xiao Liang had discussions and guidance with the students. The School of Mathematical Sciences of Peking University has done a lot of preparatory work and logistics support for this competition.

First Day
(September 21, 2020)

1 Consider the convex quadrilateral $ABCD$. The point P is in the interior of $ABCD$. The following ratio equalities hold:

$$\angle PAD : \angle PBA : \angle DPA = 1 : 2 : 3 = \angle CBP : \angle BAP : \angle BPC.$$

Prove that the following three lines meet in a point: the internal bisectors of angles $\angle ADP$ and $\angle PCB$ and the perpendicular bisector of segment AB. (Contributed by Poland)

Solution As shown in the Fig. 1.1, let $\varphi = \angle PAD, \psi = \angle CBP$. Then from the condition we know that

$$\angle PBA = 2\varphi, \ \angle DPA = 3\varphi, \ \angle BAP = 2\psi, \ \angle BPC = 3\psi.$$

Suppose X is a point on AD such that $\angle XPA = \varphi$. Thus

$$\angle PXD = \angle PAX + \angle XPA = 2\varphi = \angle DPA - \angle XPA = \angle DPA.$$

This implies that $DX = DP$, and the angle bisector of $\angle ADP$ is also the perpendicular bisector of XP. Similarly, if Y is a point on BC such that $\angle BPY = \psi$, then the angle bisector of $\angle PCB$ is also the perpendicular bisector of PY. The problem then reduces to proving that the perpendicular bisectors of XP, PY, AB are concurrent. Note that

$$\angle AXP = 180° - \angle PXD = 180° - 2\varphi = 180° - \angle PBA.$$

So $AXPB$ is a cyclic quadrilateral, and X lies on the circumcircle of $\triangle APB$. Similarly we can show that Y lies on the circumcircle of $\triangle APB$.

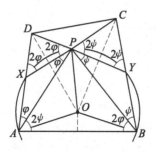

Fig. 1.1

Therefore, A, B, Y, P, X are concyclic, and the perpendicular bisectors of XP, PY, AB all pass through the center of this circle. $\qquad\square$

2 The real numbers a, b, c, d are such that $a \geq b \geq c \geq d > 0$ and $a + b + c + d = 1$. Prove that

$$(a + 2b + 3c + 4d)a^a b^b c^c d^d < 1.$$

(Contributed by Belgium)

Solution Since $a + b + c + d = 1$, it follows from the weighted AM-GM inequality that

$$a^a b^b c^c d^d \leq a \cdot a + b \cdot b + c \cdot c + d \cdot d = a^2 + b^2 + c^2 + d^2.$$

Thus, it suffices to prove that $(a + 2b + 3c + 4d)(a^2 + b^2 + c^2 + d^2) < 1$. Note that

$$(a + b + c + d)^3$$
$$= a^3 + b^3 + c^3 + d^3 + 3ab(a + b) + 3ac(a + c) + 3ad(a + d) + 3bc(b + c)$$
$$+ 3bd(b + d) + 3cd(c + d) + 6(abc + abd + acd + bcd)$$
$$> a^2(a + 3b + 3c + 3d) + b^2(3a + b + 3c + 3d) + c^2(3a + 3b + c + 3d)$$
$$+ d^2(3a + 3b + 3c + d).$$

Also, since $a \geq b \geq c \geq d > 0$ we have

$$a^2(a + 3b + 3c + 3d) + b^2(3a + b + 3c + 3d) + c^2(3a + 3b + c + 3d)$$
$$+ d^2(3a + 3b + 3c + d) \geq (a + 2b + 3c + 4d)(a^2 + b^2 + c^2 + d^2).$$

In combination, we conclude that

$$(a + b + c + d)^3 \geq (a + 2b + 3c + 4d)(a^2 + b^2 + c^2 + d^2).$$

Therefore, the original inequality is proven. $\qquad\square$

3 There are $4n$ pebbles of weights $1, 2, 3, \ldots, 4n$. Each pebble is colored in one of n colors and there are four pebbles of each color. Show that we can arrange the pebbles into two piles so that the following two conditions are both satisfied:

- The total weights of both piles are the same.
- Each pile contains two pebbles of each color.

(Contributed by Hungary)

Solution Let the n colors be $c_1, c_2, \ldots, c_n$, and let $\{1, 2, \ldots, 4n\}$ denote the set of pebbles, where i represents the pebble of weight i. We construct graph $G = (V, E)$, where $V = \{c_1, c_2, \ldots, c_n\}$, and the set of edges is defined in the following way: if pebble i has color u, and pebble $4n + 1 - i$ has color v, then we connect u, v with an edge.

Thus, G has exactly $2n$ edges, which may contain loops (which occur when i and $4n + 1 - i$ have the same color) and multi-edges. Since every color has exactly 4 pebbles, the graph G is 4-regular (note that a loop is counted 2 degrees at the same vertex). We then prove the following lemma.

Lemma *Every 4-regular graph (not necessarily simple) has a 2-regular spanning subgraph.*

Proof of lemma It suffices to prove the assertion for connected graphs, since in the general case we may take the union of the subgraphs in all the components. Let $G = (V, E)$ be 4-regular, with $|V| = n, |E| = 2n$. Then since every vertex has even degree, this graph is Eulerian and there exists an Eulerian circuit L. Now we place the edges of L alternately into E_1 and E_2, then $|E_1| = |E_2| = n$, and we claim that both graphs (V, E_1) and (V, E_2) are 2-regular.

In fact, for a vertex v, if there are no loops at v, then L passes through v twice, and every time the edge coming in and the edge going out belong to different E_i, so exactly 2 of the edges incident to v belong to E_1 and the remaining 2 belong to E_2.

If there is one loop at v, then there are 3 consecutive edges e_1, e, e_2, where e is a loop at v, and e_1, e_2 are two non-loop edges incident to v. By construction e_1, e_2 belong to the same E_i, and e belongs to the other part. Since e is counted 2 degrees, v still has degree 2 in (V, E_1) and (V, E_2).

If there are two loops at v, then since the graph is connected, the only possible case is $n = 1$. In this case, one of the two loops belongs to E_1 and the other belongs to E_2. Therefore, the same assertion still holds. Hence the lemma is proven.

Coming back to the original problem, since our graph $G = (V, E)$ is 4-regular, its edge set admits a partition E_1, E_2 with $|E_1| = |E_2| = n$, and $(V, E_1), (V, E_2)$ are both 2-regular spanning subgraphs. By the definition of G, every edge corresponds to a pair $(i, 4n + 1 - i)$. Now we put the n pairs of pebbles corresponding to the n edges in E_i into the i-th pile ($i = 1, 2$). Since each pair has total weight $4n + 1$, the total weight in each pile is exactly $n(4n + 1)$. Further, since $(V, E_1), (V, E_2)$ are 2-regular, each pile contains exactly 2 pebbles in each color.

Remark   The lemma we prove is a special case of the theorem below. For graph $G = (V, E)$, if every vertex has degree k, then we call G a k-regular graph. If a spanning subgraph H of G is r-regular, then we call H an r-factor of G. A 1-factor is also called a perfect match. **Petersen's 2-Factor Theorem.** If G is a $2k$- regular graph (that may contain loops and multi-edges), then G has a 2-factor.

Second Day
(September 22, 2020)

4   There is an integer $n > 1$. There are n^2 stations on a slope of a mountain, all at different altitudes. Each of two cable car companies, A and B, operates k cable cars; each cable car provides a transfer from one of the stations to a higher one (with no intermediate stops). The k cable cars of A have k different starting points and k different finishing points, and a cable car which starts higher also finishes higher. The same conditions hold for B. We say that two stations are linked by a company if one can start from the lower station and reach the higher one by using one or more cars of that company (no other movements between stations are allowed). Determine the smallest positive integer k for which one can guarantee that there are two stations that are linked by both companies. (Contributed by India)

Solution   The answer is $n^2 - n + 1$.

We first show that if $k \le n^2 - n$, then there exists a situation where no two stations are linked by both companies. Apparently it suffices to give an example for $k = n^2 - n$, since for smaller k we simply delete some of the cable cars.

Let $S_1, S_2, \cdots, S_{n^2}$ be the stations with increasing altitudes. We set the $n^2 - n$ cable cars of A to be from S_i to S_{i+1}, where $1 \le i \le n^2 - 1$, and $n \nmid i$. Then set the $n^2 - n$ cable cars of B to be from S_i to S_{i+n}, where $1 \le i \le n^2 - n$. Thus, two stations S_i, S_j are linked by A if and only if $\left\lceil \dfrac{i}{n} \right\rceil = \left\lceil \dfrac{j}{n} \right\rceil$, and they are linked by B if and only if $i \equiv j \pmod{n}$. These two conditions cannot be satisfied simultaneously, so no two stations are linked by both A and B.

Next, we prove that for $k = n^2 - n + 1$, such two stations always exist. Construct a directed graph G_A as follows: the vertex set consists of the n^2

stations $\{S_1, S_2, \cdots, S_{n^2}\}$, and for $1 \le i < j \le n^2$, if A runs a cable car from S_i to S_j, then there is a directed edge $S_i \to S_j$. It follows from the conditions that every vertex in G_A has at most one in-degree and at most one out-degree. Also any edge in G_A points towards the higher station, so the graph does not contain any directed cycle. It is easy to see that G_A consists of several directed chains (here a directed chain may consist of only one vertex), where two stations are connected by A if and only if they belong to the same directed chain. Since there are $n^2 - n + 1$ edges, it follows that exactly $n^2 - n + 1$ vertices have in-degree, which means exactly $n - 1$ vertices have no in-degree, so there are exactly $n - 1$ directed chains in G_A. By the pigeonhole principle, some chain contains at least $\left\lceil \dfrac{n^2}{n-1} \right\rceil = n + 2$ vertices, and let X be the set of vertices in this chain, $|X| \ge n + 2$.

Construct G_B similarly, and by the same argument G_B also has $n - 1$ directed chains. Hence, there are two vertices S_i, S_j in X that belong to the same chain in G_B, so they are linked by both companies. $\qquad\square$

5 A deck of $n > 1$ cards is given. A positive integer is written on each card. The deck has the property that the arithmetic mean of the numbers on each pair of cards is also the geometric mean of the numbers on some collection of one or more cards. For which n does it follow that the numbers on the cards are all equal? (Contributed by Estonia)

Solution For all $n > 1$, it is necessary that all cards have the same number.

Equivalently, we prove the following proposition: if $a_1 \le a_2 \le \cdots \le a_n$ are positive integers that are not all equal, then there are two numbers, whose arithmetic mean does not equal the geometric mean of one or more numbers among them.

Let $d = \gcd(a_1, \ldots, a_n)$. If $d > 1$, then we may replace $a_1, a_2, \ldots, a_n$ with $\dfrac{a_1}{d}, \dfrac{a_2}{d}, \ldots, \dfrac{a_n}{d}$. Thus, all the arithmetic and geometric means are divided by d, so the proposition remains equivalent. Hence, we may assume that $a_1, a_2, \ldots, a_n$ are coprime.

Since $a_1, a_2, \ldots, a_n$ are not all equal, we have $a_n \ge 2$. Let p be a prime factor of a_n, then there exists some $1 \le k \le n - 1$, such that $p \nmid a_k$. Among such k we choose the maximal one. Then we prove that $\dfrac{a_n + a_k}{2}$ is not the geometric mean of any part of $a_1, a_2, \ldots, a_n$.

Assume for contradiction that there exist $1 \le i_1 < \cdots < i_m \le n$ such that $\sqrt[m]{a_{i_1} a_{i_2} \cdots a_{i_m}} = \dfrac{a_k + a_n}{2}$, so that

$$2^m a_{i_1} a_{i_2} \cdots a_{i_m} = (a_k + a_n)^m.$$

If $i_m > k$, then $p | a_{i_m}$, $p \nmid a_k + a_n$, so the left-hand side of the above equation is divisible by p while the right-hand side is not, a contradiction.

If $i_m \le k$, then since $a_k < a_n$ we have

$$\sqrt[m]{a_{i_1} a_{i_2} \cdots a_{i_m}} \le a_{i_m} \le a_k < \frac{a_k + a_n}{2},$$

which is also a contradiction. Therefore, the proposition is proven.   □

6 Prove that there exists a positive constant c such that the following statement is true:

Consider an integer $n > 1$, and a set $\mathcal{S}$ of n points in the plane such that the distance between any two different points in $\mathcal{S}$ is at least 1. It follows that there is a line l separating $\mathcal{S}$, such that the distance from any point in $\mathcal{S}$ to l is at least $cn^{-1/3}$. (A line l separates a set of points $\mathcal{S}$ if some segment joining two points in $\mathcal{S}$ crosses l.)

Note. Weaker results with $cn^{-1/3}$ replaced by $cn^{-\alpha}$ may be awarded points depending on the value of the constant $\alpha > 1/3$. (Contributed by Taiwan, China)

Solution We prove that the statement is true for $c = 0.1$.

Let $\delta = cn^{-1/3}$, and for a set of points $\mathcal{S}$ in the plane and line l, we use $\delta(\mathcal{S}, l)$ to denote the minimal distance from a point in S to l. Assuming the assertion is not true, then there exists a set $\mathcal{S}$ of n points in the plane ($n \ge 2$), such that any line l separating S satisfies $\delta(\mathcal{S}, l) < \delta$. We choose two points A, B in $\mathcal{S}$, such that the distance between them is the maximal. Let $d = |AB|$. Then apparently $d \ge 1$. Now we construct a coordinate system, such that A is the origin, and $\overrightarrow{AB}$ coincides with the positive direction of the x-axis. Suppose the x-coordinates of points in S are $d_1 \le d_2 \le \cdots \le d_n$. Since the points in S are inside the intersection of the following two disks:

$$D_A = \{P \in \mathbb{R}^2 \mid |PA| \le d\}, \; D_B = \{P \in \mathbb{R}^2 \mid |PB| \le d\},$$

it follows that all the x-coordinates lie in the interval $[0, d]$, so $d_1 = 0$, $d_n = d$.

If there exists $1 \le i \le n - 1$ such that $d_{i+1} - d_i \ge 2\delta$, then line $l : x = \dfrac{d_i + d_{i+1}}{2}$ separates $\mathcal{S}$, and $\delta(\mathcal{S}, l) \ge \delta$, which is contradictory to the assumption.

Hence, we have $d_{i+1} - d_i < 2\delta$ for every $1 \leq i \leq n-1$. Now consider the points of $\mathcal{S}$ in the strip $0 \leq x \leq \dfrac{1}{2}$, and let them be $P_1, P_2, \ldots, P_k$, where P_i has coordinates (d_i, y_i), $i = 1, 2, \ldots, k$. Since $d_i \leq 2(i-1)\delta$, at least $\left\lceil \dfrac{1}{4\delta} \right\rceil$ numbers in $d_1, d_2, \ldots, d_n$ belong to interval $\left[0, \dfrac{1}{2}\right]$, i.e., $k \geq \left\lceil \dfrac{1}{4\delta} \right\rceil \geq \dfrac{1}{4\delta}$.

For $1 \leq i < j \leq k$, since $|d_j - d_i| \leq \dfrac{1}{2}$ and $|P_i P_j| \geq 1$, it follows that $|y_j - y_i| \geq \dfrac{\sqrt{3}}{2}$. Note that this implies that the difference between any two of $y_1, y_2, \ldots, y_k$ is at least $\dfrac{\sqrt{3}}{2}$, so we have

$$\max_{1 \leq i \leq k} y_i - \max_{1 \leq i \leq k} y_i \geq (k-1)\dfrac{\sqrt{3}}{2}.$$

The intersection of the strip $0 \leq x \leq \dfrac{1}{2}$ and D_B is a bow-shaped region, whose highest point and lowest point are the intersection points of $x = \dfrac{1}{2}$ and the circumference $(x-d)^2 + y^2 = d^2$, which are $\left(\dfrac{1}{2}, \pm\sqrt{d - \dfrac{1}{4}}\right)$. Consequently,

$$\max_{1 \leq i \leq k} y_i - \min_{1 \leq i \leq k} y_i \leq 2\sqrt{d - \dfrac{1}{4}} < 2\sqrt{d}.$$

In combination, together with $k \geq \dfrac{1}{4\delta} > 2$, we obtain

$$2\sqrt{d} > (k-1)\dfrac{\sqrt{3}}{2} \geq \dfrac{\sqrt{3}}{4}k \geq \dfrac{\sqrt{3}}{16\delta}.$$

Squaring both sides, and using $d < 2n\delta$, we have

$$8n\delta > 4d > \dfrac{3}{256\delta^2}.$$

Equivalently, $2048n\delta^3 > 3$. Since $\delta = cn^{-\frac{1}{3}}$, we have $2048n\delta^3 = 2048c^3 > 3$, which is not true for $c = 0.1$. Therefore, the converse assumption cannot hold, which proves our conclusion. $\qquad\square$

Printed in the United States
by Baker & Taylor Publisher Services